SMART CITIZENS, SMARTER STATE

SMART CITIZENS,
SMARTER STATE

The **Technologies** of **Expertise**
and the **Future** of **Governing**

BETH SIMONE NOVECK

▌▌▌ Harvard University Press

Cambridge, Massachusetts, and London, England • 2015

Library of Congress Cataloging-in-Publication Data

Noveck, Beth Simone.
Smart citizens, smarter state : the technologies of expertise and the future of
governing / Beth Simone Noveck.
pages cm
Includes bibliographical references and index.
ISBN 978-0-674-28605-4 (alk. paper)
1. Public administration—United States—Citizen participation.   2. Public
administration—United States—Decision making.   3. Public administration—
Technological innovations—United States.   4. Political participation—Technological
innovations—United States.   5. Political planning—United States—Citizen
participation.   6. Information society—Political aspects—United States.   7. United
States—Politics and government—21st century.   I. Title.
JK421.N29 2015
352.3'80973—dc23   2015014352

For Amedeo Max

Contents

Preface

Openness will strengthen our democracy and promote efficiency and effectiveness in Government. . . . Public engagement enhances the Government's effectiveness and improves the quality of its decisions. Knowledge is widely dispersed in society, and public officials benefit from having access to that dispersed knowledge. Executive departments and agencies should offer Americans increased opportunities to participate in policymaking and to provide their Government with the benefits of their collective expertise and information. . . . Executive departments and agencies should use innovative tools, methods, and systems to cooperate among themselves, across all levels of Government, and with nonprofit organizations, businesses, and individuals in the private sector.

—Presidential Memorandum on Transparency and Open Government, 2009

When I arrived in Washington at the start of the first Obama administration to head up the White House's "Open Government Initiative"—an effort to apply technology and innovation to improve the workings of government—I was full of optimism from having served as a volunteer adviser to the most successful Internet-based campaign in history. The Obama presidential campaign in 2007 and 2008 had encouraged its supporters to blog freely on the campaign's website. It had set up listservs and wikis dubbed "Idearaisers" to invite over five thousand experts from around the country to discuss ideas and submit them to the campaign for consideration. Early in the under-resourced campaign, volunteers on the West Coast set up campaign offices independent of the central office in Chicago. Even after the establishment of a more formal campaign apparatus, techies volunteered their time and talents to build tools to get out the vote.

Between Election Day and the inauguration, we created the first ever presidential transition website to involve the American people in the process of planning the first 100 days of the new administration. We asked people to submit both their questions and their ideas as part of a *Citizen's Briefing Book,* a unique attempt, inspired by Idearaisers, to invite citizens to tell the presidential transition team what they thought should be on the new president's first-100-day agenda. More than 125,000 people shared 44,000 ideas.

On January 21, 2009, his first day in office, as his first executive action, President Obama signed the Memorandum on Transparency and Open Government. This document committed his administration to establishing "a system of transparency, public participation, and collaboration" for government. The memorandum set in motion a national and, subsequently, an international effort to encourage public institutions to innovate to improve people's lives. The vision underlying the Open Government Memorandum was to move away from governing behind closed doors through a limited set of public policy professionals and their adjutants. The explicit goal was to move toward a world of "smarter governance" in which collaboration and conversation between government and citizens would become the default way of working on a day-to-day basis. Implementing this vision would require a radical reinvention of government: fundamentally redesigning, not merely reforming, how institutions make decisions by opening them up to the value embedded in citizens' skills, talents, and abilities.

As we unpacked our boxes in cramped White House offices, we knew that if we were going to be able to tap the intelligence and expertise of the American people and the civil service in new ways, we would need to shape our plans and craft our policies on opening government in a fashion that modeled the practices we were advocating. This was much easier said than done. We might have been on top of the world, but we were running Windows 2000. One of the newly appointed Obama administration chief technology officers of a large Cabinet department described the situation in his agency like this: "We have a nearly obsolete infrastructure. The amount of manual work is amazing, and embarrassing. Don't think Google server farm. Think gerbil on a wheel." And outdated technology infrastructure was only one of the impediments. Overzealous security mavens blocked all access to social media sites. Even the president had to exert his authority as leader of the Free World to be allowed to keep his mobile device. Statutory gift restrictions intended to prevent corruption and bribery prevented us from using even commonly used open source software. Security and accessibility inspections could delay the installation of any new software on a White House server for a year or more. The White House did not yet have a blog, let alone a Facebook or Twitter account, only an arcane and barely functioning website inherited from the Bush administration, whose idea of openness was to make a video of the White House Christmas tree from the vantage point of Barney, the First Dog. We had to contract with an organization outside government to host our initial online open government policy brainstorming sessions with the public.

When we began to address substantive priorities, we knew that a few people in a White House office and even a few hundred across federal agencies did not possess all the expertise necessary to devise policies on technical and complex subjects. But when we started to look for those people, we were quickly taught the Washington lesson that it was unwise to be "forward leaning"—that is, to tip one's hand regarding any forthcoming policies by discussing them publicly. The idea of an institution as hidebound as the White House consulting with the civil service, let alone the public, before a policy got framed and drafted was unheard of. The chief of staff of our office within the Executive Office of the President actually had to obtain the email addresses of agency heads surreptitiously in order to ask their permission for ordinary government employees to participate. The reason for these extraordinary measures: the standing practice that only the Office of Management and Budget was allowed to speak to agencies on behalf of the White House.

When we considered talking openly online about an open government policy, First Amendment concerns about government officials moderating an online forum got in the way. So did public relations concerns: for example, what if we allowed comments about UFOs or the president's birth certificate to proliferate? Such insularity was common in the White House. Admission to 1600 Pennsylvania Avenue for even a routine meeting required twenty-four-hours' notice, submission of the attendee's birth date and Social Security number, a background check, x-ray screening, and a physical inspection.

White House tours required special permission and FBI clearance. Visitors could take photographs only from a "safe" distance, behind a tall iron fence and barricades. Outsiders could not see in, and White House employees could not see out because of the voluminous synthetic bomb-blast curtains designed to catch shattered glass in the event of an explosion. Many staff members sat in Sensitive Compartmented Information Facilities (SCIFs, pronounced skiffs)—windowless workspaces required for those who routinely work on sensitive matters. Mobile phones had to be left outside the door of such rooms in open wooden cubbyholes. It looked like preschool.

Visitors to the so-called People's House—not the gilt ceremonial rooms, but our cramped, low-ceilinged offices—noticed that every desk held at least one television set broadcasting CNN, C-SPAN, Fox News, MSNBC, an internal White House channel, or a Pentagon station. Television was the buffer between the White House and regular citizens; press conferences announced policies only after they were finalized.

Public consultation took place, of course, but only after the fact, as a way to fine-tune an already formulated plan or to sell the public on it so as to ease its implementation. By the time a draft policy was presented to the public for comment, the document had typically been thoroughly reviewed by countless lawyers and policymakers within the Executive Office of the President, as well as by the leadership of multiple Cabinet agencies. Until that vetting was complete, the document was considered "deliberative" (and usually marked as such), meaning that its contents were confidential and not for public disclosure. And before surveying the public, in any case, a little-known statute called the Paperwork Reduction Act—the bête noir of the civil service—usually required a time-consuming approval process, which was, ironically, paper-intensive. Although the definition of "information collection" under this statute does not pertain to asking for public input on a draft regulation, nor does the requirement to seek approval apply when the White House (as distinct from a federal department or agency) conducts consultation, as a general matter many people avoided seeking public input for fear that they might inadvertently violate the Paperwork Reduction Act or, worse yet, have to comply with its requirements. Also, agencies have to pay a fee to publish required notices of citizen consultation in the *Federal Register,* the daily newspaper of government. So practicality dictated staying away from public engagement for reasons of cost.

When consultations were required, such as by an agency in connection with a draft regulation, good practice was to post the final—often legalistic—draft for comment after it was too late to incorporate citizen responses. By then, often only lobbyists and professional government watchers had the know-how to respond, sometimes to the end of killing a proposal.

The White House had a chief information officer responsible for tech support, and the Office of Management and Budget had a chief information officer responsible for budgetary oversight of information technology spending in the agencies. But there was no one whose job it was to develop strategies for using the Internet to attract useful contributions from citizen experts about how to accomplish major goals, such as reducing the cost of health care, improving access to education, creating jobs, and making government more efficient and effective. In fact, political leaders often invoked national security to justify shutting off access to websites and tools that could have enabled public servants to speak to—or hear from—the public. The Bush administration, for example, regularly took information about key economic indicators and terrorist threats off federal government websites. No White House had ever tried to encourage meaningful public engagement. Only a few letters sent by citizens to the White House ever make their way from the Correspondence Office to the president. Even fewer get passed along to relevant Cabinet secretaries or their staff.

We succeeded in changing some of these practices. During my tenure, the White House did a somewhat better job of broadcasting how it works and started to post online the logs of who came for meetings. During occasional video chats with senior officials the public could submit questions. But pushing information *out* did not translate into a broad willingness—or commitment—to invite more information, ideas, and voices *in*. In the midst of two wars, facing widespread dissatisfaction at home about Wall Street malfeasance, congressional bailouts, partisan backbiting, and an economy in seeming free fall, there was little appetite for more radical institutional innovation. There was virtually none for admitting that government did not have all the answers.

After the success of the initial online brainstorming process run by an outside organization on our behalf and of another conducted on a password-protected and government-only website for public servants, White House leaders briefly consented to allow blogging about our open government policy in May 2009. In essence, they chose to look the other way in allowing our use of open source software that had not been formally procured or inspected. Because of the lack of a technology infrastructure (there were no blogs used anywhere in the federal government until 2008), Brian Behlendorf, co-inventor of the Apache webserver, flew to Washington to customize a Word Press blog that we could use to discuss plans for open government and even take comments online. Although the open dialogue yielded input that was incorporated into the prescriptions of the Open Government Directive issued in December 2009 (and the federal government has added a challenge platform to run prize-backed competitions), the comment-enabled blogging platform was subsequently taken offline.

This was deeply frustrating, not least because we had become more convinced than ever that the expertise—both credentialed expertise and nonelite knowledge—existed to make government smarter. Increasingly, the challenges we face as a society are varied, complex, and momentous: confronting the prospect of a pandemic; dealing with the threats of climate change and terrorism; improving education; lowering rates of incarceration and crime; reducing domestic and campus sexual violence; addressing nuclear contamination after a Fukushima-like disaster; bringing down the billion-dollar costs of everything from healthcare to busing children to school; improving the well-being and resilience of our communities; or addressing the needs of the global poor.

At the same time, the tremendous leaps being made in science and technology could allow us to address these challenges more effectively than ever before. We can produce life-saving, portable water-sanitation kits to help eradicate disease in the most impoverished regions; invent a system of sensor technologies to monitor and improve hand-washing in hospitals and so reduce rates of death from infection; create solar cells as cheap as paint; design green buildings that produce all the energy they consume. Aydogan Ozcan of UCLA has developed a microscope attachment for a cell phone that costs less than ten dollars and transforms the device into a mobile lab to improve public health. Architect Diebedo Francis Kere has designed a thriving new school for his hometown of Gando in Burkina Faso, which has no electricity, using adobe bricks that the villager can fashion by hand without expensive machines or imported materials. There is an emerging class of socially conscious designers, architects, and technologists who are undertaking rapid experimentation with the implementation of innovative ideas for dealing with intractable social challenges. These activists embrace a citizen-centric model, working with affected people to develop site-specific solutions quickly. What we are missing—and what we lacked in the White House—are institutional designers who think about how to redesign what President John Tyler in 1840 called "the complex, but at the same time beautiful, machinery of our system of government" and the processes it uses to make decisions—and solve problems.

The design of our current institutions of governance can evolve. No law of nature (only what Roberto Unger calls institutional fetishism) holds them rigidly in place. They can be changed. Almost a century ago, American philosopher John Dewey wrote in *The Public and Its Problems,* "There is no sanctity in universal suffrage, frequent elections, majority rule, congressional and cabinet government. These things are devices evolved in the direction in which the current was moving, each wave of which involved at the time of its impulsion a minimum of departure from antecedent custom and law. The devices served a purpose; but the purpose was rather that of meeting existing needs, which had become too intense to be ignored, than that of forwarding the democratic idea. In spite of all defects, they served their own purpose well."

Indeed, the formation of states is always an experimental process of constant rediscovery of the right institutional designs for every age. No matter how difficult or improbable, change must be possible lest we end up with brittle and unyielding institutions prone to breaking when tested by complex and wicked economic and social challenges. Whether in failed states or established democracies, the goal is to shape a government that works by making better policies and by delivering services more effectively and legitimately. In our time, that requires a transformation, not just in the choice of policy but, more fundamentally, in the way we make policy in the first place—a transformation that takes seriously the capacity, intelligence, and expertise of all people and forges institutions that know how to marshal and use that human capital. Ultimately, tackling challenges from public health to educational attainment requires many people with diverse skills and talents working together. By enabling leaders and citizens to collaborate, new technologies are accelerating access not merely to information, which would only drown decisionmakers, but also to the collective knowledge and creativity needed to curate and filter innovative solutions. Yet the most ardent democrats waver in their faith that citizens possess the knowledge and the competence needed for participation in governing. We need to go beyond old-fashioned politics and ballot counting to build the conversational infrastructure that connects more diverse people and what they know to our public institutions. We need smarter governance.

The purpose of *Smarter State* is to provide both the theory and a practical blueprint for going beyond mere transparency and passive citizenship to a world in which citizenship is active and institutions are "open by default." Such openness and collaboration are essential if democratic societies are to meet the challenges of today and tomorrow.

From Open Government to Smarter Governance

> We have the power to begin the world over again.
> —Thomas Paine, *Common Sense*, 1776

Public officials have limited access to the latest, most innovative ideas in a form they can use to solve real problems.[1] The promise of opening government is that, by connecting government institutions to people and organizations with diverse forms of expertise, government will be able to produce better results.

The push for open government can be seen as a response to years of growing disenchantment with closed-door models of governance. It has become commonplace to lament the incompetence or impotence of government attempts to tackle big issues. Only 17 percent of people report having confidence in the president of the United States. Congress fares even worse. Fewer than 10 percent of people—the lowest number since surveys first began posing the question—report having confidence in Congress.[2] Silicon Valley and the technology community display their own particular brand of we-can-do-better antigovernment sentiment. According to the Roosevelt Institute's 2013 survey of a thousand young people, "While Millennials strongly believe in an activist government, fewer than 30 percent believe their voice is currently represented in the democratic process."[3]

This pandemic of distrust is not unique to the United States. Around the world, only 44 percent of people surveyed in twenty-five countries say they trust government to do what is right. Only 15 percent have a great deal of trust.[4] These perceptions that government is ineffective coincide with a growing disconnect between the potential and the practice of citizen participation. There are many knowledgeable, experienced, and well-educated people, as well as the tools to connect them to each other and to powerful sources of data and information, yet the institutions of governance persist in abjuring outside help and still underperform.

The potential for open government to restore trust in institutions is significant. Open ways of working in business (open innovation) and science (citizen science), as I shall describe, are hopeful antecedents that point to the direction democracy could take. But meaningful institutional transformation remains elusive. Previous efforts to adopt either traditional modes of citizen engagement or online participation known as crowdsourcing in any systematic fashion have failed. The lack of widespread opportunities to engage, coupled with a longstanding culture of closed-door practices, stands in the way of more open governance. Although crowdsourcing allows new groups to participate, such open call forms of engagement are still too ad hoc and unreliable to provide a safe basis for institutional reform or a blueprint for a different kind of democracy. But new *technologies of expertise* can change this equation by making new kinds of know-how manifest and searchable, which in turn makes it possible to target and match people to the right opportunities to participate. Such "smarter governance" holds the key for genuinely transforming the closed institutions of governance into open institutions that actively invite collaboration.

The core idea of open government is that governing institutions are not as *effective* or *legitimate* as they might be because they operate behind closed doors. Thus the great opportunity for them is to transition from an insular model to one in which they tap the expertise and capacity distributed in a networked society. This thinking builds on experience in the commercial sector. Many companies are enjoying great success from opening themselves up to the use of what organizational theorist Henry W. Chesbrough at Berkeley's Haas School of Business calls the "purposive inflows and outflows of knowledge" from across and outside the organization.[5] They are collaborating with customers and suppliers on the reinvention of key business practices. Such bottom-up engagement—what some call open innovation—is well known in the business literature as a significant driver of innovation.[6]

Pharmaceutical companies such as Eli Lilly started opening up their research and development (R&D) efforts in the early years of the millennium to get help from outsiders using a "solver network" called Innocentive. Synack offers companies a solver network of cyber-security experts. The "My Starbucks Idea" website asked coffee customers how to improve the company's products and services.[7] Dell Computers did something similar online with its IdeaStorm website.[8] Netflix famously gave away a million dollars to researchers who came up with ideas to improve the system it uses to recommend movies for its subscribers to watch.[9]

The evidence that companies can benefit from outside input is now well established. Many succeed by cleverly leveraging collective intelligence across distributed networks in marketplaces traditionally dominated by closed models of production. Threadless, a T-shirt company, crowdsources designs from its customers instead of employing a stable of designers. Local Motors crowdsources the design of vehicles and then sets up "microfactories" where customers can come together to collaborate on the designs and also participate in the manufacturing (think 3D printing) process. SamaSource connects people in the developing world to jobs by enabling employers to farm out tagging, research, and online data and content projects to poor workers around the world—doing well by doing good. Facebook, whose 1.44 billion active monthly users, not its ten thousand employees, create the site's content, such as forty-five billion messages sent each day, has a valuation more than four times that of traditional media companies Viacom and CBS combined, and it continues to grow in value (if not in profits).[10]

Although there are no well-developed theories of the firm that take community innovation into account, there is ample evidence that co-creation between institution and network can lead to highly relevant knowledge at low cost.[11] Studies confirm what we would expect: getting customer input and even co-creating products with customers can lead to myriad organizational benefits: original products and services, a better reputation, and useful online communities. Those who participate in these innovation processes stand to gain in various ways: from professional accomplishment or a sense of inclusivity and belonging to the pleasures of friendship, knowledge, intellectual stimulation, and fun.[12]

Technology writer Clay Shirky talks about a "cognitive surplus," by which he means the thousands of hours we spend passively watching TV that could be dedicated to creative pursuits. Today it is becoming increasingly obvious that the most valuable resource in our society is the smart citizen. We see the evidence not only in open innovation projects in the commercial world, but also in connection with the "massive outbreak" of participation in the domain of science known as citizen science. Citizen science typically refers to scientific tasks, such as data collection, measurement, and classification, undertaken collaboratively either online or off between volunteer members of the general public and professional scientists. There is a "civic surplus" too, and citizen science makes use of the enthusiasm and willingness of ordinary people to participate in measuring air or water quality in their communities, cataloging flora and fauna in museums, or analyzing satellite photographs and making maps after a natural disaster like the Nepal earthquake in 2015.

Take Galaxy Zoo, for example. There are estimated to be one hundred billion galaxies in the observable universe, each containing billions of stars. For many years, deep-view telescopes like the Hubble Space Telescope have recorded images of the Milky Way and other galaxies to help us understand how galaxies form. The resulting volume of data is enormous. After more than twenty-five years in orbit, Hubble alone has recorded more than a million observations. To begin to translate this raw information into useful scientific knowledge, the scientists at NASA have turned to "citizen scientists"—volunteer hobbyists, amateur science buffs, and space enthusiasts—to classify the images according to their shape: elliptical, spiral, lenticular, irregular. This information, in turn, illuminates the age of the galaxy.

Helping understand how galaxies form is one of dozens of such online citizen science projects. On Old Weather, volunteers do the painstaking work of retyping nineteenth-century naval shipping logs in order to create a computable database of historical weather records, which are potentially vital to unlocking the mysteries of climate change. On Old Weather, people compete to become the "captain" of a historical ship by doing more work. After it was begun in 2010, volunteers completed the work it would have taken one person twenty-eight years to do in a mere six months.

In the Open Source Drug Discovery initiative, nonprofessionals are developing life-saving advances in biomedical research. In India and around the world, thousands of primarily poor people die every day from tuberculosis. There have been no new TB treatments developed in forty years, and resistance to existing drugs is increasing. Because TB disproportionately impacts the indigent, traditional pharmaceutical companies lack the economic incentives to commit adequate resources to so-called neglected diseases. In response, Samir Brahmachari, former director general of India's Council of Scientific and Industrial Research, started the Open Source Drug Discovery project in 2008 and continues to serve as its chief mentor.

At the outset, a few thousand college students, academics, and scientists collected, annotated, and extracted data from the scientific literature on the biological properties and mechanisms of the *Mycobacterium tuberculosis* (Mtb) pathogen. Where the articles were not freely available online, students wrote to thousands of authors to request a free copy. The genome, says Brahmachari, is like a "beautifully written Shakespearean sonnet, but we don't understand the language or the meaning of it."[13] With vast quantities of data to sift through, OSDD volunteers applied the concepts of open innovation and citizen science to accelerate research and test a multitude of hypotheses. In 2014, this distributed amateur effort began clinical trials for a new experimental drug called Pretomanid.

Just as blogs turned nonprofessionals into citizen journalists, forever transforming the news industry, sites like Crowdcrafting, for performing science-related tasks that require human cognition such as image classification, transcription, and geocoding, are breeding the next generation of citizen scientists. On Crowdcrafting, participants use photographs to classify melanomas, helping researchers to better understand cancer. Ordinary people and professional scientists are making science together.

Cornell University's Project Feeder Watch employs sixteen thousand volunteer birdwatchers across North America. Digital Fishers asks citizen scientists to annotate terabytes of raw undersea video from the Neptune Canada seafloor observatory. Earthdive is a United Nations project that asks divers to record and share their experiences, including sightings of species, to create a "global snapshot of the state of our planet's oceans."

In contrast to Crowdcrafting and OSDD, which use amateurs to assist professional scientists, the Public Laboratory for Open Technology and Science (Public Lab) dubs itself a "Civic Science" project. Public Lab views citizens not as mere amanuenses to the secret and exclusionary society of scientists, but as scientists fully capable in their own right. In one project, it provides people with the tools to make maps and aerial images of environmental conditions using balloons and kites. This "grassroots mapping" project has been used to contest official maps. For instance, in Lima, Peru, members of an informal settlement developed maps of their community as evidence of their habitation. On the Gulf Coast of the United States, locally produced maps of oil spillage are being used to document damage that is underreported by company or government officials.

In science as in commerce, people are using new technology to coordinate work across physical distances and produce products without traditional market incentives or hierarchical control. The number and quality of valuable amateur peer production projects are enough to drown out any chorus of naysayers wishing to declare the notion of online collaboration naïve. It makes intuitive sense that getting ideas from outside should be just as vital for the performance of public institutions as it is for private institutions.[14] Neither can innovate in isolation. As Friedrich Hayek wrote in 1945, "The knowledge of the circumstances of which we must make use never exists in concentrated or integrated form."[15] Networks enable institutions to gain access to the wealth of creativity and insight that exists in the wider society.[16]

Given the plummeting rates of trust in government, the appeal of such ideas is no surprise. Following scandals both sexual and financial, even the Catholic Church turned to the laity for input in the rewriting of Church doctrine.[17] Clearly there is widespread enthusiasm for the potential of open innovation. As President Obama declared at the launch of the Open Government Partnership at the United Nations in 2011: "Put simply, our countries are stronger when we engage citizens beyond the halls of government. This, I believe, is how progress will be achieved in the 21st century."[18] Similarly, European Parliamentarians have stated that "the public must have the opportunity to participate in policy-making,"[19] and Prime Minister David Cameron has argued "the best way to ensure an economy delivers long term success for all its people is to have it overseen by political institutions in which everyone can share."[20]

Open government, pronounced East African civil rights leader and philanthropist Rakesh Rajani, "is the most apt response to the democratic human impulse to be involved, to count, to matter." We want this kind of change in government because technology has transformed the material conditions of social life leading to "a massive cognitive and emotional transformation," according to political commentator Moises Naïm, "which is a growing dissatisfaction with political systems and government institutions, public and private."[21]

The Potential Benefits of Open Government

To understand why working "openly" in government, not only in commerce or science, could improve how we produce public goods and manage the allocation of scarce resources, consider how a more open government might address the challenge of reducing recidivism among prisoners.[22] A few facts: the past two decades have seen a dramatic reduction in the number of prison-based college education programs, from roughly 350 in 1995 to 12 in 2005.[23] This reduction has come at a cost. Studies show that college classes for prisoners reduce recidivism significantly.[24] There is an economic justification for doing more, too, since the average cost of housing an inmate for one year is almost $32,000, nearly six times the size of a Pell Grant, the funding awarded to needy students to subsidize their college education. But spending on such programs is at the same level that it was thirty years ago, even though spending on corrections has steadily increased over the same period. One of the few remaining college education programs for incarcerated felons, Boston University's College Behind Bars, reports that, of the hundreds of prisoners who have taken its college classes and participated in a parallel volunteer mentoring program, only two have returned to prison.[25]

Tackling the issue of recidivism would benefit from outside input at several stages. First, understanding the problem requires an empirical understanding of the status quo, which could be improved by sharing the data government holds about the prison system with computer scientists outside government and then inviting them to create the visualizations, models, and tools to help everyone better understand the problem. We need data scientists to document the baseline of current data sharing in the criminal justice system and to identify key criminal justice information. Legal scholars and practitioners, prisoner and victim rights groups, criminologists, and policemen, as well as prisoners and their families, are in a position to provide qualitative input about the realities of conditions on the ground.

After the problem is understood, practical solutions for how to deliver the right kind of program that will reduce recidivism and crime cheaply still have to be identified. Here, engaging those with relevant experiences, skills, or insights in, for example, online education, vocational training, criminology, psychology, mental health, and process reengineering would add value. Those experts might sit inside government agencies like the Department of Education or Department of Health as well as within universities or companies. Entrepreneurs will have compelling ideas about how to get iPads into the hands of prisoners to assess learning ability and skill and deliver GED-style education customized to the learner.

After an approach is identified, then social scientists could play an important role by designing pilot projects and experiments to test what works, by comparing one intervention with another. Business leaders and philanthropists might prove useful at translating policy into practical implementation by creating training and job opportunities, online and off. With greater outside input, finding solutions would put less onus on government to devise new kinds of public-private partnership arrangements. After the fact, those with training in evaluation and assessment could help figure out what works, comparing interventions, and evolving solutions informed by real-time data. In a world of finite resources, why would we ever ignore a bounty of rich and diverse insights, assistance, or collaboration?

Opening up how we govern has the potential to yield *more and better insights* than those generated by government working alone. The federal government, for example, contracts with airlines to provide travel to federal employees. Left unmanaged, these exclusive arrangements would allow airlines to charge the highest possible prices. But by sharing government operations data about travel and its costs, the General Services Administration was able to sponsor a "travel data challenge" in spring 2014 and invite the public to spot better ways to save the public's money.

Getting new insights and ideas for how to solve problems, big and small, was the core of the idea behind Challenge.gov, the federal government's challenge platform. Since its launch in 2010, Challenge.gov has been used by over seventy federal agencies to crowdsource solutions and the site has been visited 3.5 million times by people in 220 countries and territories and more than 11,000 U.S. cities. It has hosted over 400 scientific, engineering, design, multimedia, ideation, and software challenges, from predicting the severity of the flu season to designing the best 3D printed Christmas ornament. It has also encouraged many imitators at the state and local level. In San Francisco since 2012, ImproveSF has engaged the public in helping to solve civic challenges, from designing a new library card to making public parks safer.

Even in the smallest towns, citizens are sharing insights. For example, the city of Manor, Texas (pop. 5,235), bet that asking the public how to save money would generate creative new ideas. Using its own open innovation platform, town officials invited the public to devise ways to do more with less. In 2010, citizens identified a grant for which the town was eligible and collaborated on drafting a successful proposal that funded putting computers in cop cars (before there were iPhones) to improve the efficacy of policing. As Manor's former chief information officer, Dustin Haisler, explains, citizen engagement is a service of government as essential as providing water or sanitation.[26] And in cash-strapped communities, the potential for collaboration and co-creation with citizens to save money is paramount.

These insights can also come from *greater scrutiny*. "More eyeballs" will more easily spot waste, fraud, and abuse in a national or local budget. This happened throughout the first decade of the millennium, when India's leading freedom of information group, Mazdoor Kisan Shakti Sangathan (MKSS), which translates as the Association for the Empowerment of Workers and Peasants, invited villagers throughout the state of Rajasthan to meet in their town square to review data painted on a wall (or conveyed by song, street play, or poster)—what they call social audits—in an effort to spot evidence that dead people were on the government payroll.[27] Beginning in 2012, it also happened when citizens in the leafy London suburb of Redbridge examined a visualization of the town's budget on a website.[28] Now that same data-publishing tool from Redbridge is being used across England to consult with citizens about their local budgets. And Kaggle enables organizations (companies more than governments) to use the crowdsourcing insights of top data scientists around the world to help with fraud prevention, while DrivenData has a similar business model, but focused on public and societal clients.[29]

In addition to stimulating new ideas, governing openly also holds the promise of finding *more diverse approaches* to solving problems. Consider the United Kingdom's National Health Service (NHS) Institute for Innovation and Improvement. Its goal was to generate and implement ideas that improve patient experiences from, for, and by patients themselves rather than just health care professionals. In its first competition, which ended in 2013, nine projects ultimately shared a £1 million grant and all were implemented. More diverse participation leads to more robust, nuanced, and complex approaches—and to greater awareness. Citizen reporting of problems through such systems as AMBER Alert, storm trackers, earthquake damage reports, forest fire alarms, and electronic suggestion boxes on government websites take advantage of people's unique situational awareness.[30]

Diversity also implies participation by *new actors* who have never engaged with government before. Brainstorming, online comments, mashathons (where developers put multiple datasets together to create a useful tool in a weekend), hackathons (where coders quickly develop a new app in a similarly short timeframe), data dives (events where data scientists come together to tackle a social problem quickly), datapaloozas (which are much the same but generally include a more diverse mix of industry, academia, and activists), and new forms of peer network engagement where people come together online and off with the civil service, often informed by data, are attracting people who do not typically engage with government. When India asked the public to help redesign the rupee in 2010, it did not ask people to ratify or even choose among designs developed by the government. It asked them to submit their own designs. The winning entry came from a graduate student.

Governing more openly also gets *more hands on deck*. Patrick Meier is currently the director of Social Innovation at the Qatar Foundation's Computing Research Institute. His Micromappers project enables "anyone with an Internet connection and five minutes to spare" to contribute to disaster relief, greatly expanding the resources available in the event of a tragedy.[31] The toolkit enables his large network of "digital Jedi" to sift through quantities of text, photo, and video information—what Meier calls Big Crisis Data—and label it quickly in order to make sense of a volume of information during and after a disaster. All over Latin America, people are volunteering to help map the instances of public health outbreaks, as in Honduras Health Mapping, or to map crime in Mexico City on the Mapa Delectivo.[32]

In Chennai, India, a city of four million people, there are more than 5,000 separate bus routes. The official bus map was incomplete and incomprehensible and had been "under construction" for six years. So in 2010, Arun Ganesh, a student at the National Institute of Design in Bangalore, decided to design a new map. He turned to the Web to request help from his fellow bus passengers. Commuters began contributing timetables and bus details, and in just three days he had compiled enough data to create a fresh map for mobile phones with a clean, comprehensible design. Using the data, he then developed a visualization of where the bus network was failing to provide adequate service.

Opening government also has the potential to bring *more resources to bear* on key challenges. Crowdfunding, where people contribute small amounts of money rather than time or attention, allows the public sector to *do more with less* in the face of decreasing taxpayer budgets. U.S. Census Bureau data, for example, show that 2011 was the first time per-student spending declined nationally in public education, at least since data collection began in 1977. In Texas, funding dropped in real dollars over the first decade of the 2000s. With no other options, the School at St. George Place, a public school in Houston, used crowdfunding to work around its budget shortfall. Using DonorsChoose.org, a crowdfunding platform that matches donors to schools, the school raised over $43,000 in the following year for supplies. More controversially, residents of Oakland, California, crowdfunded a private police force to supplement what they saw as inadequate government services. Crowdfunding is also a mechanism to build support for projects people care about that might not otherwise be implemented. Fresh from municipal bankruptcy and constrained by a court-mandated spending plan, Central Falls, Rhode Island, a former mill town, raised $10,000 in 2014 to beautify and clean up a landmark park. Although civic crowdfunding often only raises small amounts, it is used to attract bigger sources of support.

Finally, working more openly can enable the *refinement and improvement* of policies at all stages of the policymaking process. Refining a proposal is the basis for the Liquid Feedback platform, which encourages Pirate Party members in Germany and Five Star Movement members in Italy to work together on policy documents for their respective political parties.[33] Podemos in Spain is utilizing a popular platform from Argentina called DemocracyOS and a host of additional online tools to the same end. Plataforma Brasil is another suite of original authoring and annotation tools created specifically for "crowdlaw," namely online legislative drafting purposes.

Openness also encourages those inside government to improve how they work as a result of the *audience effect*. Knowing that someone is listening and watching often leads to better performance by those inside institutions, especially when they are doing their usual jobs.[34]

The Failure of Open Government

The White House Open Government Initiative—and the Memorandum on Open Government, with its tripartite manifesto of transparency, participation, and collaboration—can boast some successes since its launch in 2009. It has inspired many of the examples just described. It has also led to one important policy shift, namely, a movement to open up government data. The memorandum declared that public information must be disclosed rapidly "in forms that the public can readily find and use." To be considered open, such data had to be freely reusable without significant legal or technological restrictions. The idea that data collected and held by the government, about society and the economy as well as about the workings of government, should be available to everyone has earned some currency.

The initiative helped bring about the creation of data.gov, a national open government website designed to make more data readily findable. Today, the governments of more than forty countries have made more than a million data sets available online for free public use and re-use.[35] These include data sets about the workings of government, such as budgets and data about government spending on contracts and grants. But open data is much broader than data about the workings of government, such as politicians' tax returns, who-met-with-whom, and even spending data.[36] Rather, it includes any data that can be publicly released that government collects about the economy, the environment, and the marketplace.

For example, the Canadian Open Government Directive calls for releasing government "information and data of business value to support transparency, accountability, citizen engagement, and socioeconomic benefits through reuse, subject to applicable restrictions associated with privacy, confidentiality, and security."[37] The United States Open Government Directive, the policy which directed federal agencies in how to implement the Open Government Memorandum, defines the data to be opened as "high value data," which is any information "to increase agency accountability and responsiveness; improve public knowledge of the agency and its operations; further the core mission of the agency; create economic opportunity; or respond to need and demand as identified through public consultation."[38]

The last decade or so has seen a data explosion around the world. Until recently, much of this data was closed to public access—hidden behind firewalls and passwords, protected by copyright or other legal barriers. In addition, some of the data was kept from the public due to legitimate concerns over privacy or national security. Recent years have witnessed something of a sea change in the way data is treated, particularly by governments and, to a lesser extent, private companies. Across the world, there has been a gradual—and sometimes not so gradual—transition underway from what McKinsey calls a culture of "protect" to one of "share, with protections."[39] It is this move toward greater data transparency that marks what some observers have called an open data revolution. Where Big Data is the information science trend of being able to collect, transmit, visualize, and mash up ever larger quantities of information, Open Data refers to the policy of making data available such that it can be readily accessed, used, and re-distributed free of charge, often via a one-stop data sharing platform.[40] The technology of Big Data fits hand in glove with the policy of Open Data. More significantly, people outside of government are using the data to develop solutions to problems. When governments around the world make subway and bus schedules available as open data, civic coders are able to create smartphone apps that tell commuters when their bus or train is coming.

The Open Government Initiative in the United States sparked international collaboration among over sixty countries from Mexico to Croatia to Indonesia and even Russia, which appointed a cabinet-level minister of open government. Each of these nations has become a signatory to the Open Government Partnership (OGP), which commits them to rethink how they govern. Membership in the partnership requires completing the projects outlined in each country's open government plan. Most of the participating countries are making some effort to share data, such as government spending information, online.

But many of the plans are also characterized by recycled policy initiatives, meaningless lip service to the notion of engagement, and open-ended promises to make changes without any commitment to measure their impact. And some suffer under the stigma of inconsistent government action. Critics complain that Russia enacted an open government plan but then invaded Ukraine (Russia subsequently withdrew from the OGP). In the wake of the Snowden revelations, the United States' leadership on open government was lambasted for what many perceived to be duplicity and hypocrisy.

The Philippines, Finland, Italy, and Brazil, among others, have conducted experiments to engage directly with citizens in developing policy and legislation online.[41] The Icelandic government asked citizens to contribute to drafting a new constitution.[42] These emerging "crowdlaw" projects (which encompass the drafting of regulations, laws, and constitutions) all share a common goal: to develop workable and replicable frameworks for providing people with opportunities to participate directly in the legislative process. But these are still one-offs and do not yet add up to a systematic effort, in either the United States or elsewhere, to change the default approach in policymaking from closed to open.

Community-led participatory budgeting, whereby the public plays a role in identifying, discussing, and prioritizing local public spending projects, is still the primary example of governance pushing out power to citizens. Fifteen hundred local legislatures, from New York City to Seville, Spain, to Porto Alegre, Brazil, are experimenting with handing control over millions of budget dollars to citizens, not professional politicians and officials. But participatory budgeting remains extremely localized and largely limited to simple capital spending projects, such as fixing a school or building a soccer pitch.

The current welter of "open gov" activity points to a *potential* shift from a top-down, closed approach to more decentralized, collaborative, and open practices of solving problems and creating public goods that take the intelligence and expertise—the smarts—of the public seriously. Yet there is little evidence that systematic change is under way, and little understanding of what ultimately works.

Examples of tech-fueled governance innovation such as open data sets, crowdsourcing platforms, hackathons, and participatory budgeting are becoming more commonplace, but high barriers between citizens and the professionals who administer and legislate remain the norm. Despite notable advances in the use of technology in elections, especially large-scale datasets to improve get-out-the-vote and fundraising efforts, technology has changed very little about how elected officials actually govern. Governing is the still the domain of professionals working behind closed doors.

For those who do not take full-time jobs in government or in quasi-governmental federally funded R&D centers, or who are not handpicked for one of a few advisory committees that meet in Washington, DC, a few times a year, there is no obvious channel for participation. The Department of Health and Human Services cannot easily consult with a network of doctors, nurses, or public health professionals to plan a new health care policy. There is no simple mechanism for a team of scientists to help the Environmental Protection Agency evaluate climate change data. There is no easy way for an economist and a physicist who want to help the Treasury (or a governor or a mayor) to collaborate on modeling economic forecasts. A new mayor in America has no systematic way to find out how to increase broadband deployment in his city. A research scientist who has ideas about how to improve the efficiency and quality of public housing will find no way to offer her services or test her ideas. The leader of a student initiative to measure air quality across campuses or send college students into public schools to teach science will find no blueprint for how to connect student manpower to public funding. Except by voting every four years, citizens have few formal ways to contribute to the improvement of civic institutions and programs.

Indeed, despite all the enthusiasm for and widespread recognition of the potential benefits of more open governance, the open government movement has had *remarkably little effect* on how we make public decisions, solve problems, and allocate public goods. Although technology has upended every other industry, including politics and campaigning, it has barely affected how the tools of actual governance—legislation, regulation, service delivery, taxation, spending, dispute resolution—are wielded.

In 2011, diverse protests by the Occupy movement (launched in New York City's Zuccotti Park) and the Twitter-fueled Arab Spring movement, encompassed demands to change how policy is made. Their ideas spread rapidly around the globe. "The process of bottom-up direct democracy," wrote Nathan Schneider, who chronicled the

#OccupyWallStreet movement for *Harper's,* "would be the occupa-
tion's chief message, not some call for legislation to be passed from
on high."[43] The medium for protest was also the message. Occupy
noticeably lacked managerial control of the kind that political parties, unions, bureaucracies, and companies have depended on.[44] To
the surprise of most, these protest movements were nevertheless effective at capturing the public imagination, demonstrating that, as
journalist Jeff Jarvis observed, "one need no longer control institutions to control agendas."[45]

But after the spring comes the fall. Technologies like Twitter and
Facebook were well suited to complaining about government, expressing moral outrage, sharing information, and coordinating protests, both physical and virtual. They did not and do not provide the
infrastructure for governing itself. It is true that online protests in
2012 against U.S. congressional attempts to regulate the Internet, as
well as against a major breast cancer foundation that had announced
its intention to cease subsidizing cancer screenings conducted by
Planned Parenthood, undoubtedly led to an about-face in policy. Text
messaging, photo sharing, and social networking did help to defeat
bad proposals. But they have been much less successful at *creating*
good new policies.[46]

For instance, the small San Francisco nonprofit Invisible Children
created in spring 2012 the Internet's fastest ever word-of-mouth campaign, against the Ugandan warlord Joseph Kony, who abducted
children to serve as soldiers and sex slaves. A video it released online
received more than 100 million viewers in one week. But nothing
changed as a result.

When the Boko Haram terrorist group escalated its ongoing campaign of terror in Nigeria by kidnapping three hundred schoolgirls
in April 2014, people marveled at the spontaneous outpouring of in-
dignation that erupted on Twitter under the banner #bringbackour-
girls, arguably forcing Nigeria's president to accept outside assistance
with a search-and-rescue mission. But as of this writing, only some
of girls have been found and Boko Haram's attacks have continued.

Despite numerous individual examples of success with open government, we are failing to govern differently the *day after* the Arab Spring. Social media activists have made effective use of technology to marshal supporters but have yet to learn how to convert that energy from challenging power into changing it.[47] Open government continues to be largely aspirational.

The Limits of Participatory Bureaucracy

Bureaucrats do, of course, converse with the public in order to build supportive coalitions, gather information, and facilitate policy implementation. The conventional practices for doing so include town hall meetings, advisory committees, and notice-and-comment rulemaking. But traditional public participation practices—or what Susan Moffitt calls "participatory bureaucracy"—do not put a chink in the armor of bureaucratic administration. These "intrusions of democracy on bureaucracy" engage only small numbers of people in infrequent conversations.[48] Official rulemaking provides only a narrow avenue through which to solicit expertise. Since World War II, the Federal Administrative Procedure Act has required that agencies publish pending regulations for public comment—but only after they are drafted. Many believe these processes are inadequate and reflect, not the outcome of genuine participation and deliberation, but political wrangling between entrenched interest groups.[49] Even the consultations that do take place have become "routinized and pro forma" negotiations between lobbyists and bureaucrats, as Francis Fukuyama contends. Ex-post comments on draft laws and regulations come too late and are often in a form that is not useful to decisionmakers.[50]

Advisory committees can provide technical expertise to agencies, but often such committees are made up of scientists employed by industry rather than academic institutions or public interest organizations.[51] Much has been written about the deficits of these committees, in particular their lack of balanced political views.[52] Even when industry views don't dominate, however, advisory committees are slow to form, have little diversity, and are constrained by law and practice to represent stakeholders rather than to provide expertise. Advisory committees *advise*.[53] They have no ongoing decision-making or operational authority. They convene infrequently and are not well integrated into regular problem-solving or policymaking practices.

Think tanks are no better. When they first began to appear at the turn of the twentieth century, they reflected a Progressive Era belief that expertise from the social sciences could be brought to bear to solve social problems.[54] Business interests then also weighed in in favor of positive, scientific, and objective standards for government that would ameliorate the social dislocation caused by industrialization and provide firms with the greater predictability they needed to achieve commercial success. Hence it was not only Taylorism—the early twentieth century science of management advocated by Frederick Winslow Taylor, which believed that measurable advances in efficiency and productivity could be achieved—that fueled the interest of business in professionalism in government; it was also a desire to prevent social unrest among the working poor against business interests.

Whatever their original role as information brokers helping to get ideas to policymakers in useful formats, think tanks have changed dramatically since their inception. These institutions of expertise increasingly pursue aggressive campaigns with a specific ideological point of view. In a study of contemporary think tanks, Andrew Rich explains that, despite their nonprofit status, they are active participants in partisan battles. Through aggressive marketing campaigns, they are "focused more on providing skewed commentary than neutral analysis," often at the end of the policymaking process, when sides have already been chosen. If anything, says Rich, think tanks have all but "neutralized the power of expertise in American policymaking."[55] Moreover, foreign governments and entities have paid tens of millions of dollars in recent years to think tanks to sponsor "research" that promotes their interests. Although this is obviously lobbying by another name, these activities usually go unregistered. For example, the Center for Global Development, a nonprofit research organization, received $5 million from the Norwegian Ministry of Foreign Affairs to push government officials to double their aid spending.[56]

Federally funded research and development centers (FFRDCs) such as Rand and Mitre, and national laboratories such as Sandia and Lawrence Livermore, are essentially extensions of federal agencies, created after World War II to expand the hiring of technical and scientific professionals in response to perceived security threats. As public-private partnerships, FFRDCs are able to collaborate closely with government behind the firewall and are free from procurement restrictions. At the same time, they enjoy the flexibility of operating as private sector institutions. But the FFRDCs, which are creatures of the cold war, currently serve primarily the Defense Department and the Department of Energy, which has responsibility for the nation's nuclear stockpile. Like a federal agency, they employ full-time professionals and are no better-suited to offering agile, real-time, or diverse advice on a range of topics.

The Limits of the Open Call

In 2005, my students and I began building a crowdsourcing platform called Peer to Patent that would allow volunteer scientists and technologists to share their expertise with the U.S. Patent and Trademark Office (PTO). At the time, the PTO had almost a million pending applications. With only fifteen to twenty hours to examine each one and decide if the inventor deserved a twenty-year grant of monopoly rights, patent examiners needed faster access to the information relevant to their decisionmaking. Although skilled in the legal art of patent examination, examiners did not necessarily have the scientific training to evaluate whether a piece of software or a business process deserved a patent because it was new or represented a significant advancement over what preceded it. Peer to Patent provided a way for scientists and technology experts working in industry and academia to share what they knew about the claims of a patent application. To ensure that the Patent Office received a manageable amount of the most relevant and well-vetted feedback, Peer to Patent enabled people to rate and rank each other's contributions.[57]

The quality of information it provided was fine, but Peer to Patent had a hard time *attracting* participants. At best, it drew three or four dozen volunteers to share what they knew about the science in a single patent application. More often, the numbers were closer to three or four, if that. We found that getting even a few of the right people to participate required extensive "marketing" of the opportunity via emails to friends, contacts, and connections. We did research in the blogosphere—this was pre-Twitter—to see who knew about the technology at issue and invited those bloggers and their readers to get involved. We looked for people with the insights and experience, and with passion for the specific scientific and technological issues at stake in each patent application and then told them about the opportunity to engage. And then we had to do it over again and again and again.

What we found is a very common issue in public decisionmaking: the lack of clear, cost-effective, and reliable ways to find those with the right expertise, insights, information, and innovative solutions. One study in the UK found that, even with software tools at their disposal, a mere 27 percent of public sector employees said they were able to locate people with the expertise they needed.[58]

On any given issue, the most interested and invested stakeholders are not necessarily experts. In the patent domain, for example, those who are most interested in patent law, such as intellectual property lawyers and lobbyists, are not scientists. And those who know the relevant science often do not know how to navigate the Patent Office bureaucracy. What a crowdsourcing platform like Peer to Patent demonstrated, against widespread skepticism, was that the public could offer useful information and would do so constructively. It showed in practice that "openness" could lead to transformative improvement. But Peer to Patent was, at best, only a partial solution. Our "open call" approach, typical for crowdsourcing, was to announce the opportunity to participate and hope people would join in. It was not reliable. It did not allow us to identify easily subject matter experts or target them effectively. This was not unusual. Government, like most large organizations, has a hard time getting to know its audience.

In some contexts, crowdsourcing is a fabulous way to get better ideas faster. When it works—when a heretofore unknown individual or group within a crowd proposes something startling and enlightening—it is a legitimate cause for celebration. During the upheaval after the Kenyan elections in 2007, the blogger Ory Okolloh wrote, "It occurs to me that it will be useful to keep a record of [the violence]. . . . Guys looking to do something—any techies out there willing to do a mashup of where the violence and destruction is occurring using Google Maps?"[59] By *luck,* a developer living in Kenya read the post and forwarded it to a developer of Kenyan origin in the United States. Together they took up the call and built the first version of what became the life-saving platform Ushahidi.org. *Ushahidi* is the Swahili word for testimony, and the platform they developed has been deployed in Kenya to track the violence; it has been reused in other crises to create collaborative maps of forest fires in Russia and Italy, as well as to pinpoint the locations of victims under the earthquake rubble in Haiti.

A compelling idea can be persuasive, but the people most able to help are often unknown to those who have a problem to solve and are therefore hard to find. They can come from unlikely places. An Argentinian auto mechanic extrapolated from the trick of using a plastic bag to extract a cork that had fallen into a wine bottle to invent a revolutionary, life-saving device to extract babies from the birth canal—the first real advance over the forceps in four hundred years.[60] His good idea eventually came to the attention of the World Health Organization, but only because he shared it on YouTube. Who would ever have imagined?

These successes are random and serendipitous. For every open call that works, there are dozens that are never seen by those who could help. Although good luck makes for wonderful stories, it does not make for reliable governing.

There have been efforts to complement crowdsourcing with the offer of a prize or other extrinsic incentive. Prizes, after all, create an incentive for insightful enthusiasts to emerge from the woodwork.[61] When the Air Force needed to solve the problem of how to halt an uncooperative vehicle at a checkpoint without hurting bystanders, it issued a sixty-day prize-backed challenge and posted the contest on Challenge.gov. (Police departments face the same challenge. Uzis work well in movies, but not in real life.) A thousand people showed interest, and a hundred submitted solutions. The winner was a 65-year-old retired mechanical engineer from Lima, Peru, who proposed a remote-controlled robotic vehicle that catches up to the fleeing vehicle and deploys an airbag underneath it, lifting it off the ground and bringing it to a stop. The Peruvian won $20,000, and the Air Force has developed an electro-mechanical prototype to test the concept prior to production.

As budgets tighten and information and communication technologies continue to advance, prize-backed challenges, contests, and grand challenges offer the advantage of paying only for solutions that work and widening the pool of potential problem solvers beyond the "usual suspects."[62] Posing a challenge backed by a cash or reputational prize can attract a large number of potential solvers with good solutions to a hard problem. The Xprize, which heavily markets its large-cash-reward grand challenges, has brought a great deal of attention to the role of cash incentives in spurring innovation. Among other things, it has challenged people to develop a 100-mile-per-gallon vehicle and a handheld "Tricorder" (think *Star Trek*) that can monitor and diagnose medical conditions. These grand challenges offer very large rewards for measurable and specific advances in tackling compelling, often scientific, problems, such as making solar energy as cheap as coal, or electric vehicles as affordable as gas-powered ones, or identifying all asteroid threats to humankind and an approach to overcoming them. They are based on "ambitious goals on a national or global scale that capture the imagination and demand advances in innovation and breakthroughs in science and technology."[63] The success of prizes has already led to legislative changes in the United States that legalize their use in the public sector at the national level.[64]

But prize-backed challenges rely on having a big enough haystack and an enticing enough monetary purse to coax out the desired needle. Otherwise, "the chance of finding that person," commented Alok Das, senior research scientist at the Air Force Labs who ran the vehicle challenge, "is zero."[65] Prizes work. Social media and prizes accelerate and increase the opportunities to make connections. They are, however, not fast, systematic, or reliable enough to affect markedly how government makes most decisions or solves most problems. We cannot replace the Department of Education with a series of prizes.

If those who are most likely to be able to help do not hear about the problem, the opportunity for engagement is squandered. Participation is random. This is true even when a prize accompanies the call. Crowdsourcing widely does not transform the epistemic ability of institutions to obtain and use innovations. Governments and organizations have not been able to find all those retired engineers or to match them with specific opportunities where they can help. We cannot and should not dismantle longstanding institutions because of Twitter. None of the tools we have created or deployed in the public sector can yet enable institutions to *systematically* target expertise wherever it lives. The high costs of finding the right expertise— insights we can rely on—means we have to continue to rely on professionals for governing, and they prefer to operate behind closed doors.

That is why many organizations, from investment and law firms to sports teams and nonprofits, do not rely on the serendipity of crowdsourcing. Instead, they go to great lengths to *target* expertise systematically. Specialty consultants, such as those who procure expert witnesses, maintain an old-fashioned Rolodex of advisers and match them to the needs of firms. Sports teams have a network of scouts whose job is to spot talent. Some investors never spend a dollar without first consulting with the experts procured for them by a bespoke expertise finder such as the Gerson Lehrman Group in New York City. Organizations such as the MacArthur Foundation rely on a vetted network of people to identify and recommend extraordinary, unorthodox, and game-changing ideas and individuals for their fellowships. Now firms are turning increasingly to hi-tech solutions such as the London-based Profinda, which mines a company's emails and documents to help it get smarter about its own expertise.

As successful as crowdsourcing has been in the years since Jeff Howe first put a name to online engagement in *Wired* magazine, with or without prizes, it will never be the basis of sustained institutional innovation.[66] Relying on serendipity ultimately reinforces the tendency to continue making policy in traditional ways. Whether they are identifying problems, brainstorming solutions, drafting policies and proposals, or implementing strategies, officials facing the uncertainty and high cost of finding experts with the appropriate experience, skills, abilities, and interests will continue to rely on internal professionals. They cannot depend on *crowdsourcing widely* when what is needed is to *crowdsource smartly*.

Why Closed-Door Governing Remains Rational

The persistence of closed government in the face of developments in crowdsourcing, citizen science, and open data is not the result of some malevolent power grab. Change has been slow to come because, for all its faults, professionalized government, as an institutional arrangement, remains highly efficient and rational. Professionals still matter. In our environment of information overload, human intelligence is still required to structure the efficient flow of information in order to produce public goods and decisions about social and economic life. Aligning information with decisionmaking requires expertise because only experts can curate what is relevant, specific, and credible. The professions have long served as a proxy for expertise, helping to identify quickly who knows what. Because professionals must conform to certain rules and standards for categorizing and organizing information, they offer a reliable and legitimate source of authority for decisionmaking.

In an administrative system, where an elite group of leaders is supposed to solve critical problems largely on their own, without really engaging the insights, experience, and brainpower of the people they serve, the professionals are the experts. They move in circles socially distant from the lives of everyday citizens and speak in esoteric and jargon-filled tongues that reinforce their elite position.[67] Robert Dahl speaks of the "quasi-guardianship" of the public by policy specialists.[68] The citizen is just a spectator—or what Michael Schudson calls a monitorial citizen, who ratifies or rejects decisions made by others.[69] The citizen's job is to watch (and hold accountable) the professionals as they debate and decide.

Because we conflate professional with expert, and define expertise—and hence power—in terms of credentials rather than skills or experience, we are not able to take advantage of the expertise possessed by those outside of government. That is, in designing our institutions and the political philosophy on which are they are based, we have set up a false dichotomy between "public"—understood as the unwashed masses—and "experts," who are professionals either inside government or part of a small credentialed circle close to government. It is true that most people know little and care less about politics. But it is also true that we now can perceive clearly that what Harold Laski called the "plain man" possesses extraordinary knowhow, skills, experience, and passions—the broader palette of human capacity. At the same time, we must break free of what William Blake called our "mind forged manacles"—our old, ironclad assumption that only professionals possess the necessary expertise to govern well.

Going "open" is, therefore, a call to exercise civic muscles that have atrophied. It requires acknowledging that regular people possess deep knowledge, wide experience, and significant training related to the solution of real problems. When we demean the value of peer-based projects—for example, crowds transcribing the works of Jeremy Bentham or ancient Egyptian papyri—as *amateur* efforts, we damn them with the faintest praise.[70] Delightful stories about "ordinary" outsiders making science, music, and art can obscure the fact that, although amateurs might not be professionals, they can contribute to tackling public problems and possess a passion to do so. This is why Bill and Melinda Gates announced a commitment in their annual letter for 2015 to support the creation of a global data-base of citizens, where people can register their interests in helping to fight global poverty.[71] Especially in response to a global crisis like a pandemic, we need lists of trained personnel who can be mobilized, says Gates.

Such ideas run counter to the thoroughgoing professionalization of public life that emerged during the industrial revolution, when the tools for measuring and ordering the otherwise unpredictable natural world—such as clocks and watches to regulate time, the steam engine to regulate energy, standard weights to regulate measurement, and canning to regulate safe food availability—enabled the transforma-tion from premodern institutions based on patronage and privateering to professional bureaucracies and public management rooted in sci-entific positivism. Skilled, salaried civil servants could manage the production of public goods more efficiently than had been possible before.[72]

Professionalization began in America as a result of industrializa-tion, scientific positivism, and the need to create a secure source of middle-class jobs in the civil service. There have always been doctors, lawyers, and architects. Law, however, was long a vocational subject taught by practitioners in offices or training schools, not a scientific discipline taught in universities. During the late nineteenth century, craft knowledge about everything from funerals to philanthropy became "scientized," as did public administration. Professionalism, industrialization, and universities burgeoned in tandem, feeding and catalyzing each other's success. Professional societies of doctors, lawyers, bankers, and public managers—with their journals, jargon, codes of conduct, and internally legally binding rules—created an ever-wider chasm between members and nonmembers. The result: in the public sector today, government professionals are treated as having a monopoly on competence. The doors of governance are closed, even to professionals in other domains, and propped shut by highly complex, formalistic, "official" processes for decisionmaking. They demand an almost anthropological familiarity with native political lingo and are bolstered by legal restrictions that prevent outside engagement.

Professionalization has created three additional cultural hurdles that stand in the way of opening up our governing institutions. These impediments are even more established than partisanship and politicking. First is the "myth of spectator citizenship"—the belief, even among reformers, that only professional public servants possess the requisite skills and abilities to govern. In this view, which pervades both democratic political theory and the practices of policymakers, citizens may express opinions based on values, but never expertise rooted in scientific facts: asking people for their opinions is one thing, but whether out of cognitive incapacity or laziness, the general public will not participate productively; expertise is possessed only by elites with professional credentials. By blinding ourselves to the expertise that ordinary people possess, we have reduced the role of citizen to that of watcher. If we do not acknowledge the ability of the public to engage productively, it is no wonder that smarter governance is slow to arrive.

Second, public policy professionals have cultivated a decision-making culture that is top-down in orientation, unnecessarily complex, and lacks the mindset or skillset for experimentation. The twentieth-century bureaucratic state has fostered mechanistic, cause-and-effect approaches rooted in the faith that professionals trained in the science of governing can discover the "right" solutions to virtually any problem. With the best and brightest professionals around the table, the thinking goes, we will find the right answer. But our institutional systems are broken. Our adherence to formalistic processes is not catalyzing the desired outcomes. What results instead, writes Geoff Mulgan, are "energy systems designed to produce and distribute energy, but not to use it well; a food system that generates worsening obesity; a health system dominated by hospitals ill-suited to populations suffering from long-term conditions, including mental illness; social care systems wholly unprepared for rapid aging; and, among others, economic systems that still suffer profound imbalances of unused resources and unmet needs. In each case, these systems that now look broken are as much victims of success as anything else. As so often, success repeated itself until it finally became failure, which is why those at the heart of each of these systems are often the last to understand how they need to change."[73]

We have come to understand that in the complex systems we inhabit we cannot intervene with any predictability in the course of events and that what is needed instead are more innovative and evolutionary approaches to problems. We are on the brink of a new era in which institutions will need to be better configured to respond to the unfolding of complex world events in the network economy. But at present we are still wedded to a professional ideology of mechanistic, expert social management with which open government is incompatible.

Third, and perhaps most lacking, are tangible "mental models" for knowing what the alternative would be to professionalized governance. The widest chasm between today's closed door and tomorrow's more conversational models of governing is a shared understanding of the goal toward which we must work. A mental model is a representation of how something works. Ironically, one of the inventors of the term explained it precisely as the thing we do not have about government. Having a mental model—a concrete and specific picture of smarter institutions—is a precondition for convincing the large numbers of people who must be involved in such a transition that it is worthwhile. In his book *Changing Minds,* Howard Gardener explains that changing people's perceptions is not a process of creating epiphanies but of catalyzing gradual shifts through examples and practical learning.

The desire for stable governance—propped up by old mental paradigms, institutional resistance to change, and a lack of funds to invest in governance innovation—helps ensure the persistence of outdated institutional models. Nothing fundamental stands in the way of a concerted effort to build a genuinely smarter state—nothing, that is, except the immense dead weight of entrenched institutional arrangements and beliefs that have long served us well but that now slow us down: in professionalism, in spectator citizenship, in mechanistic institutions, and in the sufficiency of existing mental models.

True, the Internet enables more distributed forms of information and communication sharing. But an open call for engagement (what we nowadays call crowdsourcing) cannot be the basis for governance transformation. It is too unreliable as a mechanism for ensuring an adequate flow of information into and out of institutions. It is too haphazard. Professionals, whose credentials reduce the cost of searching for and finding relevant expertise, are quite good curators and brokers of ideas. In many instances, relying on the crowd without first curating and identifying key participants, results only in the same old ideas being proposed and rehashed.[74] Like well-known consumer brands, traditional professional credentials are labels that reduce the cost of finding what we are looking for. New technologies are changing this equation. Just as brands were once excellent proxies for choosing between products, so too were the professions and their credentials an excellent proxy for locating trustworthy expertise.

Although turning all policy operations over to professional, university-trained government civil servants may have made sense at one point in history—indeed, it was a key assumption of twentieth-century progressivism—it no longer does. To be clear, the problem is not professionalism per se but one of exclusionary and exclusive practices that limit participation and collaboration with those outside government, including other elites and credentialed professionals as well as those with practical know-how. Changes in technology and

in social organization made possible by technology allow governments to harvest information and assistance from a diverse array of
experts and use this information to promote better policy decisionmaking and more effective delivery of public services. The tools are
better. Now we have to use them.

The Consequences of Our Failure to Innovate

In the United States, our institutions of governance are largely functional and effective. Most of our public policy professionals are just
that—unstintingly professional and capable. But failing to open up
how we make policy already imposes significant opportunity costs.
Complex challenges often seem intractable because the governing
practices we use are neither robust enough to incorporate the full diversity of potential approaches, nor sufficiently attuned to people's
social nature.

Because these challenges are massive in scope and impact, they
require us to think differently about the way we run our organizations. Consider the Food and Drug Administration (FDA), which
manages the complex process of premarket approval and postmarket
review for all medical devices. The pathway to regulatory review for
low-risk items like tongue depressors is straightforward; but more
complex, life-sustaining, and high-risk devices—for example, bioelec-
tronic edible batteries, which offer a dissolvable power source for
new kinds of ingestible medical devices—require judicious premarket
approval by those with the right know-how.[75] And they require it
quickly if they are going to save as many lives as possible. If these devices reach patients before being properly tested, the human costs
could be high. If review is too slow and cumbersome, delays in getting to market could cost both lives and jobs. Because medical devices
are steadily growing more complex and increasingly involve new sciences, from nanotech to 3D printing, it can take up to nine months
simply to convene a qualified review panel. If the FDA continues to

rely on a directory of the same old people, these processes will take longer and longer without getting better.

By contrast, in 2009, the United States Patent and Trademark Office opened up the data about its own workings and began to embrace, albeit in a limited fashion, public participation in patent review and other innovations in how it works. The result: the queue for review is declining and quality is increasing. Open and agile institutions have repeatedly shown they can get smarter faster.

But open practices are not simply an accelerant; they are also a mechanism for breaking the cycle of path dependence when wrong choices get made too early. Government policies are blunt instruments that inevitably suffer from unintended consequences, which are often large enough to nullify their positive effects.[76] Moreover, in today's budgetary climate, it is no longer cheap and easy to reverse course when we make the wrong decision. Of course this means we need innovative institutions that bring the best science and technology to bear on governance choices. But as decisions get harder, we need to make them in open and collaborative ways that enable us to test what works and then adjust when necessary.

Consider what has happened with military uniforms. In 2002, every serviceman in the United States had a choice between green and "coffee stained" camouflage. Then, over the next decade or so, each of the armed services decided to create its own fashion-forward design without conducting any research about these designs' potential to protect lives or prevent injury. The Army alone spent $3.2 million designing and another $5 billion producing "universal" camouflage that was supposed to blend in anywhere, but works exactly nowhere. And then it shelled out millions more to replace the replacement camouflage. The cost of fixing this mess across the military is estimated to be another $4 billion. We should have long ago embraced evidence-based decisionmaking and moved on from following what retired lieutenant colonel and camouflage expert Timothy O'Neill describes as the "CDI"—"chicks dig it"—factor in making such decisions.

Like most important policy choices, however, keeping soldiers safer is not an exact science solvable only with data. It cannot be *solved;* it can only be *managed.* Hence the answer cannot be practices based on ego or hierarchy. Nor can it be simply to bring credentialed people into government—the progressive illusion that the "best and brightest" technocrats can fix any problem.[77] Rather, the solution is to make decisions by openly testing, critiquing, and advancing—in conversation with those who possess both scientific and practical know-how—our understanding of what works.

In contrast to the uniform debacle, the health care reform legislation passed in the United States in 2013 embraced openness and experimentation. As Atul Gawande explains in the *New Yorker,* half of the legislation is devoted to programs that would "test various ways to curb costs and increase quality."[78] It is a hodgepodge of pilot programs and experiments designed to reduce the staggering costs and declining quality of a health care system that is imperiling our economy. The law even created a center devoted to generating and testing experimental innovations. The jury is still out as to the success of this approach, but early results are promising.

Old-fashioned, slavish adherence to unbending rules and calcified practices—as if there were only one right way to do things because things have always been done that way in government—virtually guarantees that we will not be able to tackle the big, systemic issues of the future. Such backward-looking, elitist practices also constrain us by forcing us to accept the professional's "God's eye" view of important problems—that is, looking down on them from the Olympian heights of Washington, Brussels, or other bureaucratic centers of power to which only the rich and powerful have access.[79] In a much-talked-about study, political scientists Martin Gilens of Princeton and Benjamin Page of Northwestern found that the preferences of rich people had a much bigger impact on subsequent policy decisions than the views of middle-income and poor Americans. Indeed, the opinions of lower-income groups, as well as of the interest groups that represent them, appear to have little or no independent impact on policy.[80] Worse, these elitist ways of working do not provide the tangible, ground-level, and human-scale insight that keep us from crafting inequitable, unjust, and unfair policies or offering poorly designed and managed services that, by failing to put the citizen at the center, put an already frayed social fabric further at risk.

In Queens, New York, for example, there have been an estimated 100,000 illegal apartment conversions—often shoddy, unsafe, sublet basement rooms and apartments rented to illegal immigrants. In one recent year, building inspectors failed to gain entrance to these properties 67 percent of the 23,410 times they tried. Even of the 8,345 units about which they had received specific complaints, they could not conduct an inspection 39 percent of the time. (To enter over an owner's objections, inspectors must have a warrant, which they requested for less than 1 percent of the properties where they had been refused permission to enter.) In 2008, a fatal fire swept through one of these basement apartments, killing three men. The apartment had been illegally split into four one-room dwellings with only one exit—a major violation of the building code.[81] When the government fails to keep citizens—especially vulnerable citizens—safe at even the most basic level, the social fabric frays even more.

When New York City opened up its building inspection process in 2012 and started using data provided *by citizens* who called the 311 hotline with complaints about illegal apartments, the city's new chief analytics officer found something startling. He discovered that the illegal conversions were creating circumstances that led to the death or injury of firefighters, who often rushed into such burning basements at their own peril. The fire department was then easily persuaded to accompany the building inspectors on inspections. Fire marshals cannot be turned away when they knock.

Or take as an example the real-world problem of hunger in America. One in six Americans lives in a household that cannot afford adequate food. Fifty million people are "food insecure," and 17 million of those individuals skip at least one meal a day to get by. Being food insecure means having to make daily trade-offs between paying for food, paying the mortgage, or paying for medical care. When choosing what to eat, many of these individuals must make choices between lesser quantities of higher-quality food and more abundant but less-nutrient-dense processed foods, the consumption of which leads to more expensive health care problems down the road. The same amount of money buys one-tenth the calories in fresh fruits and vegetables than in energy-dense but low-value junk food. The challenge is not made any easier when poorly designed public assistance programs continue to count the sauce on a pizza as a vegetable. The deficiencies caused by hunger increase the likelihood that a child will drop out of school, lowering her lifetime earning potential. Millions of people depend on assistance from state and private food-aid programs, and millions more need but do not get any help.

The crisis will not disappear without government intervention. But it also will not end with one right policy; this scourge cannot be "solved." Indeed, the root causes of poverty are complicated and highly interdependent, necessitating myriad interventions to combat it from different angles. As much as we need specific policy innovations to address food insecurity, we also need institutions capable of identifying and implementing innovations.

We cannot crowdsource an end to hunger. Open and collaborative institutions engaged in ongoing conversations with both credentialed experts and citizens with situational awareness could, however, improve our ability to ameliorate hunger in America. Identifying who is genuinely in need cannot be done as well by a centralized government bureaucracy—even with regional offices—as it can through a distributed network of individuals and organizations pinpointing where the demand is greatest. Streamlining the administration of services reduces costs as well as the indignity of applying for benefits. Redesigning such processes, however, demands both a new institutional mindset and assistance from those outside government with experience and expertise in human-centered design.

With greater citizen engagement and oversight, possibly facilitated through online platforms, it also becomes more difficult for Congress to cave to the pressures of lobbying groups that push for subsidies for crops—such as white potatoes and corn—that are primary ingredients in less-nutritious foods.

Creative ideas from outside government can improve the quality of the services delivered, too. There are exciting experiments using incentives to nudge people toward purchasing more healthful food. New kinds of public-private partnerships can create the means for people to produce their own food. Both new kinds of financing arrangements and new online applications (apps) for managing the shared use of common real estate could make more community gardens possible. And with the kind of attention, convening, and funding that government can bring to an issue, new neighbor-helping-neighbor programs—where people take turns shopping and cooking for one another to save time—could be expanded. Measuring what is working and what is not can be done far better by an army of citizen scientists reporting data through an app than by a small number of government inspectors.

There is no irresistible law of nature that says democratic institutions must grow stiff, rigid, and unresponsive. Yet in comparison with the vast amount of research devoted to politics, political dysfunction, and political civility, and to business and science innovation, there is surprisingly little attention being paid to *institutional innovation*. There is far too much reliance on "anecdotes and hunches" than there is to understanding the processes by which institutions might use information differently and work more collaboratively to improve people's lives.[82] This is not the same as asking how to get more science into decisionmaking. Instead, the focus should be on how we can tackle important public problems and how science and technology can transform the way we do so.[83]

Why Smarter Governance May Be the Answer

Despite the promise of what open government could do to improve both the legitimacy and the effectiveness of governing, the persistence of traditional closed-door governing has made perfect sense. Many talented people would participate constructively, but it has not been practical to find them and pair them with opportunities that match their abilities and interests. As a result, engagement has been fitful and unreliable. The public sector in many places like Mexico and the United States has expanded fellowship programs to bring talented people into government to do public service. As Mikey Dickerson, ex-Googler and head of the United States Government Digital Service, explains his choice to return permanently to the public sector after fixing the healthcare.gov website: "It was . . . more important and meaningful than anything I could have accomplished in a lifetime working at my old job."[84] We have also developed a set of practices for after-the-fact citizen engagement, such as commenting on rule-making, rooted more in the belief that outside participation is a nice-to-have for the sake of legitimacy but not a must-have for active citizenship or effective governing. Even with the introduction of online tools for crowdsourcing and online collaboration, there is still no systematic way to increase engagement.

Whereas crowdsourcing and engagement platforms make it possible to collaborate with people online, expert networks make "smarter governance" possible, namely "targeting" and matching people with whom to collaborate. These technologies of expertise are enabling individuals, groups, and teams to express a broad range of talents, skills, and abilities digitally. New online course platforms are democratizing how we learn; new badging tools are decentralizing the accreditation of know-how; new data science techniques are making it possible to discover individuals' know-how, not just their publications or grants or citations; and search tools such as LinkedIn are making people with a wide range of skills findable online. Whether it is a badge from an online learning platform such as Khan Academy that attests to my mastery of a skill, or a customer rating on a tutoring platform such as WyzAnt or Helpouts that attests to my ability to teach that skill, these new tools divorce the concept of "expertise" from institutional credentials and automate the discovery of expertise discovery both within and across organizations. By culling and collating information about people's credentials, as well as their experiences, skills, and interests, these tools use a combination of big data from across the Web and information entered manually by people about themselves and others to produce information-rich ways of profiling, searching for, and finding expertise.

Take, for example, PulsePoint, a software application created by the Fire Department of San Ramon, California, to enable citizens to assist with first response in medical emergencies. Unlike an open call, PulsePoint does not ask anyone to participate in a mass effort. Rather, it targets trained volunteers with knowledge of cardiopulmonary resuscitation (CPR). Each year, 424,000 people in the United States suffer sudden cardiac arrest and roughly a thousand die each day. Effective CPR administered immediately after a cardiac arrest can double or triple the victim's chance of survival, but less than half of victims receive that immediate help. A bystander can do three things to improve a victim's chances for survival: call 911, start chest compressions, and/or use a defibrillator. If a bystander who knows how to do these things arrives on the scene of an emergency in the first few minutes, the odds of death go down 50 percent. According to the American Heart Association, bystander CPR is performed only one-quarter of the time. PulsePoint aims to change this. PulsePoint enables local 911 emergency services to notify *registered* and *trained* CPR users—off-duty doctors, nurses, police, and trained amateurs, for example—to come to the aid of stricken neighbors. PulsePoint sends them a text message in capital letters: "CPR Needed." To date, PulsePoint has activated over 11,000 citizen responders across 1,100 communities in the United States.

The GoodSAM App in the UK (SAM stands for Smartphone Activated Medics), a similar service that evolved from London's Air

Ambulance, targets an even narrower category of doctors, nurses, paramedics, and other off-duty professionals. To be approved as a responder, volunteers must first upload their work ID, such as a hospital badge along with their professional registration number.

In PulsePoint's first five years of operation, eleven thousand nearby citizen rescuers came to the aid of four thousand heart attack victims in more than eleven hundred communities across fourteen states. As services like PulsePoint and GoodSAM demonstrate, people are smart and capable; they have specific skills and abilities; and if they are aware of an urgent need, many are willing to help.

By cataloging who knows what, these searchable directories could help to unlock three different types of hard-to-get expertise. First, they make it possible to uncover more about the skills, both theoretical and practical, of professionals working *inside government*, where bureaucratic titles disclose little. Rather than knowing only that someone is an assistant deputy director, imagine seeing a list of the projects he or she has worked on and with whom, the rules they have drafted or reviewed, the policies they have helped to create and implement, as well as practical skills in data scraping, human-centric design, contracting and procurement, or customer service. Whether credentialed or uncredentialed, research-based or practical, we know far too little about the human capital inside public institutions.

Using these tools, leaders in government could make better use of the abilities of lower-ranked civil servants. The New York Police Department, for example, maintains a database of its employees' abilities, from languages spoken like Hausa or Hindi, to special skills such as beekeeping experience or music, in the event that there is a call to wrangle runaway bees from an urban hive in Brooklyn or to sing the national anthem in Chinese at a school graduation in Chinatown. Members of the service share what they know by completing a form as part of the human resources intake process. Common requests for specialty skills include demand for those with a pilot's license for the Emergency Service Unit or diving certification for the Harbor Unit, as well as searches for those with information technology training.[85] Although little data is kept about either the supply or the demand for expertise and how the police make use of it, the Department takes the task seriously enough to track hundreds of possible skills from acupuncture to yoga instruction.

Second, such tools are also making it possible to find credentialed expertise *outside* government. Imagine being able to connect policymakers with bright professors and professionals outside of government. Only a handful of people can serve on advisory committees, but many more have deep research-based expertise to share. With limited pathways available for academics and other professionals to participate in policymaking, many universities are using research networking platforms to catalog the expertise of their faculties. VIVO, for example, is a publicly funded project for making biomedical researchers searchable within and across universities. The Conversation, a blogging platform for professors, offers a searchable directory of their academic bloggers organized by area of expertise.

Many credentialed experts are not academics. But as successful professionals, they may be participants in social networks that provide access to useful problem solvers. However, because these experts are not inside government or universities, their knowledge has been especially difficult to find and harvest.

Third, the technologies of expertise may be unlocking *practical proficiencies*, such as the ability to perform CPR. These tools can help identify people with horticultural expertise (through their participation in online forums); people with skill certifications earned in online classes or situational awareness from having lived in a particular location; people with practical experience making, doing, building, and inventing; and people with special kinds of acumen and insight, whether physical or mathematical.

With the diffusion of such tools, expertise is no longer exclusively the domain of elites with particular social attributes; it is a neutral description of different forms of knowledge within a much broader population. That expertise might be elite and university-credentialed or non-elite and non-credentialed; it might be used to describe those inside or outside of government; and it might be an ascription of theoretical or practical know how. In every case, technology has the potential to make that expertise more searchable and useful. This is not to say that participation in this vision of smarter governance is a free-for-all. Unlike with voting, the goal is not for everyone to participate, but to increase the number, variety, and frequency of opportunities to engage. Targeting specific kinds of people will, indeed, establish hierarchies of talent and ability; different people will be called upon more in certain circumstances. Just as most Wikipedia readers never edit an online article, and most open source software users never write a line of code, so, too, participation will not become universal.

But targeting expertise has the potential to open up participation to much larger swaths of people both inside and outside government, creating cross-cutting and alternative forms of access. These technologies hold the key to transforming closed institutions into open ones that actively invite those with relevant skills and interests into the conversation.

Without the ability to match people to problems, relying on professionals working behind closed doors was the most rational way to decrease the costs of finding the expertise needed to govern. But today more people than ever are getting smarter without professional training or university credentials, and they are willing to share their expert knowledge and undertake complex tasks if asked. The opportunity for people to present a more nuanced and diverse picture of their achievements and abilities can open up new pathways for learning and advancement. But, more significantly from the perspective of governing, these tools make it possible to pinpoint those with a broad range of useful skills. Work samples, citation and download data, peer and customer reviews, scores and badges, and leaderboards are making abilities and aptitude searchable with a level of granularity that is not obvious from degrees, diplomas, and transcripts.

So, if a city really wants to improve the chances of crafting a workable plan for bike lanes, it should be able to reach out to urban plan-

ners, transportation engineers, cyclists, and cab drivers and offer them
ways to participate meaningfully. If it wants to address questions of
hospital performance, it should have a way to reach out to doctors,
nurses, hospital workers, patients, and their families. Doing so does
not exclude anyone else who wants to participate. It also does not pre-
clude civic engagement between citizens independent of government. Rather, knowing who knows what enhances the ability to alert
those who would most like a chance to collaborate in governing and
to apply their talents in the public interest. The technologies of expertise point the way toward a radical transformation in how we run
our organizations. When we make expertise of all kinds systematically findable, participation has the potential to become robust and
commonplace, and citizenship has the potential to become more active
and meaningful.

The Rise of Professional Government

Professions are conspiracies against the laity.
—George Bernard Shaw, *The Doctor's Dilemma*, 1906

George Washington. Thomas Jefferson. James Madison. Alexander Hamilton. Everyone knows the famous Founding Fathers. But fewer have heard of Ebenezer Bowman or Stephen Chambers of Pennsylvania, Oliver Ellsworth of Connecticut, or Nathaniel Gorham of Massachusetts, despite the important roles they played. They are just a few of the thousands—the "we the people"—who fought for and ratified the U.S. Constitution in each state's constitutional convention. Despite the complexity of the task, the drafters of the Constitution in Philadelphia took the unprecedented step of calling for ratification by the people themselves, not by the Confederation Congress or even by the state legislatures—that is, by amateurs not professionals.

Tens of thousands of ordinary people fought the Revolutionary War, not just what historian T. H. Breen called "a handful of elite gentlemen arguing about political theory." So when John Adams later moved into the still unfinished executive mansion in Washington, he knew that the very people with whom he had worked shoulder to shoulder against the British had sent him there. More than just a residence, this whitewashed, sandstone house almost immediately came to symbolize a special and fragile democracy. Nosy neighbors snooped with impunity around the construction site. In 1801, Jefferson threw open the doors to visitors and installed exhibits of exotic animals for popular entertainment. He instructed Lewis and Clark to offer his hospitality to the Native American tribes they encountered during their explorations of America's vast open spaces, and soon chiefs from several Indian nations were camped on the lawn of the "Great Chief" in Washington. Once, when John Quincy Adams was meeting with Secretary of State Henry Clay, one Eleazar Parraly wandered in to say hello and shake the president's hand. Parraly introduced himself as a local dentist. The president immediately dismissed the secretary of state and asked Parraly to remove an aching tooth. Abraham Lincoln admitted soldiers to the first floor where they occasionally took naps on the sofas. Franklin Pierce, who became president in the 1850s, once remarked to a passerby who asked to come in and see the fine house: "Why, my dear sir, that is not my house. It is the people's house! You shall certainly go through it if you wish."

By present-day standards, informality reigned. Despite Hamilton's ardent hope for an efficient administrative apparatus, before the late nineteenth century, there was no formal professional civil service in America. To be clear, the absence of a professional public service and extensive constitutional provisions about the administrative state did not mean there was not already plenty of administration. The levying of taxes, collecting of duties on ships, and establishment of a private patent system, for example, were of vital importance to generate revenue for the debt-ridden fledgling nation. Doling out pensions to veterans and their widows and the provisioning of relief to those suffering from disasters were among the welfare state initiatives of the early nineteenth century. Imposing Jefferson's embargo on all foreign ships between 1807 and 1809 was an important foreign policy maneuver, but implementing it was a herculean feat of administrative prowess.[1]

Nonetheless, administration was a patchwork of practices, mostly local in nature. Merit was not considered the key factor for employment. There were no political science departments to train those who worked in the government. Indeed, the social sciences did not yet exist as an organized academic field. Neither was there any stabilized, cross-cutting administrative law.

The practice and study of government lay not in the writings of scholars but in "citizens' literature."[2] Its documents were not textbooks. Instead they were items like those included in the syllabus taught by the University of Virginia's first professor of law and civil polity: the Constitution, the Federalist Papers, the Farewell Address of George Washington, the decisions of Chief Justice John Marshall, and the Resolutions of the General Assemblies of Virginia and Kentucky.[3] The Founding Fathers were not professionals but men who blended the "Gentleman with the Frontiersman to make the new American Commonman."[4]

In the new communities of the wide-open Northwest, ordinary people participated actively in public affairs. Frederick Jackson Turner extolled the amateur governance of the American forest and frontier in his seminal book *The Significance of the Frontier in American History*: "Every militia muster, every cabin-raising, scow-launching, shooting match, and logrolling was in itself a political assembly where leading figures of the neighborhood made speeches, read certificates, and contended for votes."[5] The roster of offices to be filled and operated was blank, and men of no previous political experience had to do the job.

By the middle of the nineteenth century, however, the Age of the Founders had passed. With it died the idea of citizens' literature and gentleman lawmen and laissez-faire frontier democracy. It was replaced at the national and even at the local level by professionals who decided and governed. Lost were both the informality of access and the robustness of informal participation that characterized life before industrialization, when everyone had to do his part in running things. As the nineteenth century progressed, the country grew, and technology advanced. Progressive concerns about urban problems—labor and social welfare, municipal reform, and consumer protection—and agrarian interests in railroad, tariff, and trust regulation took shape in opposition to the excesses of the emerging capitalist economy.[6] In turn, these fueled a demand for stronger government. Today, in contrast with the 780 government employees (excluding deputy postmasters) in 1792, there are more than two million civilian "professional" public servants employed in the federal Executive Branch.[7]

So where did this modern notion of administration as almost exclusively the province of professionals come from? How did it become so deeply rooted in our political culture that today we take it for granted? Why do we no longer engage the public at large in the governance of public institutions? Why is authority now the province of lawyers and public policy experts? There is no way, of course, to do justice to the administrative history of the United States during the nineteenth century in a few pages. But if, today, there is a case to be made for again taking citizen expertise seriously and for making the intelligence of the public a foundation for institutional innovation, then we need to understand why day-to-day citizen participation and capacity fell out of favor.

The professionalization of public administration in the late nineteenth and early twentieth centuries was designed to apply scientific principles to governing while reducing corruption, patronage, and inefficiency. Inadvertently, it excluded the public from meaningful participation in governance. "It may be accepted as a fact, however unpleasant," lamented Theodore Roosevelt, "that if steady work and much attention to detail are required, ordinary citizens, to whom participation in politics is merely a disagreeable duty, will always be beaten by the organized army of politicians to whom it is both business, duty, and pleasure."[8] This exclusion shaped—and was shaped by—a belief that people are either unwilling or unable to contribute to their own governance except through partisan politics and the pursuit of individual interests.

Reliance on professionals to govern developed in parallel to three social trends that became dominant in the mid-nineteenth century: industrialization and the emergence of organizations to accommodate the scale and complexity of modern life; standardization of weights, measures, and tools for controlling social conditions; and the rise of professional training in universities. These three trends, along with the emergent ideology of professionalism, did much to shape the modern design of political institutions and public decisionmaking. They made professionalism key to the legitimacy of government actions, and public participation the exception rather than the rule.

The Roots of Professionalism

Professions, by their very nature, exclude those who do not belong to them from sharing in the body of knowledge their members claim. Although the definition of professionalism has been the subject of intense debates within the academy, the concept has a common core, which emerged in the latter half of the twentieth century, when writing about the professions peaked.[9] Sociologists who study it (political scientists do not) offer fairly similar definitions of its central characteristics: work rooted in a unique, cognitive domain of learning and knowledge.[10] Put another way, professionalism refers to a group's knowledge-based competence.[11] Their specialized training, outside the workplace, most often in universities, produces the credentials to distinguish professional from layman. The more standardized this training, the stronger the profession. The less competitive the market for the services or products produced by a professional group, the stronger the profession. The broader the mix of people served and the greater the affinity of the profession's ideology with the surrounding culture's dominant ideology, the stronger the profession.

Although professions clearly enjoy economic advantages as a result of their exclusive and exalted status, there is also something noneconomic that points, instead, to the exercise of a sense of public duty and purpose, which breeds deference. That is why we usually accede to the professional's competence and commitment to act responsibly. Such monopoly over a domain of knowledge easily translates into power and authority over those who depend on that knowledge. "To most of us," writes Paul Starr, "this power seems legitimate: When professionals claim to be authoritative about the nature of reality, whether it is the structure of the atom, the ego, or the universe, we generally defer to their judgment."[12] But why?

Professionals are organized bodies of experts whose knowledge appears to the rest of us to be esoteric. They have elaborate systems of instruction and training. They possess and enforce a code of ethics or behavior more strict than the one the rest of us follow.[13] They work hard to maintain their exclusivity through private licensing associations sanctioned by the state, which affords their societies significant latitude to exclude others and to regulate and even censor members' speech like a state actor.[14] The more such bodies assert some kind of political control and exclusivity, the more quickly they graduate from mere occupations into professions.[15] The American Medical Association and the American Bar Association, for example, regard such deference as essential to their independence and competence.

Professions impose and police the epistemic rules for cataloging how certain types of knowledge are produced and disseminated within their ranks and, as a result, wield tremendous political power. This capacity to arrange experiences in useful and predictable ways, comments Robert Post, dean of the Yale Law School, is central to the process upon which a robust and thriving democracy depends.[16] Unlike law or medicine or architecture, however, which must rely on the state to delegate to professional organizations the right to license practitioners and thus to control and manage the boundaries of the profession, civil service professionals can ensure the boundaries of their own profession and shore up the authority of their legitimate domination, as well as their ability to exclude others, by developing legal strictures that they enforce using esoteric language that prohibits outside engagement in policymaking activities.

The role of the professional civil servant is enshrined by the law itself, which reinforces the profession's control over the flow of information into and out of institutions—what Pierre Bourdieu calls the "officializing strategy" of bureaucracy—in ways designed to dissuade citizens from engagement.[17] There is a wealth of administrative law that limits control over speech in the public sector to public management professionals and treats their decisionmaking with legal deference. For example, key information law statutes intentionally limit information sharing and collaboration and preserve the domain of the public servant distinct from and closed to others. The public earned a right to access information held by government relatively late in the twentieth century, and even then only upon request and with significant limitations. For those of us outside the curtain, the effect is impressive.

Today the policymaking profession is, for the most part, split among professionals trained in law, political science, and the newer fields of public administration and public policy, which were founded closer to the middle of the twentieth century to unify the production of policymakers and scholarly knowledge about policymaking.[18]

Just as some other professions were able to restrict the supply of practitioners, the civil service was able to confer an economic advantage on its members.[19] These bureaucrats, educated by modern universities in the positive, rational principles of scientific management, created a stable and merit-based set of job opportunities for a growing middle class. Both Tocqueville and Chevalier remarked on the absence of both a proletarian and an aristocratic class in nineteenth-century America, where the middle classes, especially lawyers, held sway. From that middle class emerged the dominant paradigm of the trained professional. As William J. Goode wrote in 1960, "an industrializing society is a professionalizing society."[20] Between 1800 and 1900, the term "professional" doubled in use in published works; a hundred years later, the usage had grown more than eightfold. It is now a firm and well-protected part of the institutional landscape.

This rise of a professional bureaucracy parallels the development of the technologies of accurate measurement, which civil servants use to manage and control society. Before the mid-nineteenth century, it was a major challenge to measure things accurately: time, distances, the creation and distribution of reliable power. Before the industrial age, wind powered the sails of ships. An animal pulled ploughs. Water turned waterwheels. There were innovations, of course. Waterwheels were lined up to coordinate output; cranks and gears allowed power to be transferred further; the Brest wheel and the turbine increased productivity; and standard weights and measures came to replace diverse local measures. But when the water dried up with changes in weather and seasons, power diminished. Power was not only limited and low, but also unreliable.[21] Work was limited by the cycles of the setting sun. And it was impossible to know with certainty what time it was or where one was located (hence the handsome Longitude Prize proffered in 1714 to reward the discovery of a strategy for ascertaining the location of a ship at sea). There was simply no sure and consistent means of ordering life in ways we take for granted today. Without the ability to synchronize clocks, one could not be paid for an hour's work or rely on people being on time or showing up at all.

In 1790, 95 percent of the U.S. population lived on scattered farms and in small towns and villages of a few hundred. In 1800, New York City was a town of only sixty thousand. The tenth largest city in America was Norfolk, Virginia, which had a population of six thousand. (A hundred years later, Cincinnati was the tenth largest, with 325,000 people, and New York had become a metropolis of three and a half million.) As semi-subsistence farmers, most Americans were highly susceptible to the instabilities of nature. The conditions of life were extremely variable—and unpredictable.

"The pre-modern state," says anthropologist James Scott, "was partially blind; it knew precious little about its subjects, their wealth, their landholdings and yields, their location, their very identity. It lacked anything like a detailed 'map' of its terrain and its people. It lacked, for the most part, a measure, a metric, that would allow it to 'translate' what it knew into a common standard necessary for a synoptic view."[22] As a result of this blindness, the flow of information between governed and government was fitful and imperfect. The attempt to collect taxes, for example, was highly uneven, with most communities only paying a fraction of what they owed, if that. All kinds of special assessments, as well as military campaigns and other methods, arose for extracting rents in the face of government's inability to control or predict revenues.

Premodern society necessarily developed institutions to help it cope with life given the technological limitations of the day. Before industrialization, two ubiquitous political institutions helped to manage the unreliability of social order: patronage and private incentives. Trust between patron and servant (that is, say, between the Earl of Sandwich and Samuel Pepys) was a primary form of social ordering in pre-industrial England and elsewhere. Aristocratic hierarchy, its mores and rituals, provided the means to create trust and manage social relations.[23] The British Navy, preeminent in its day, was an aristocratic institution run by an officer corps appointed through a complicated patronage system. Offices such as military commissions and tax collectorships, widely viewed as venal institutions, were privatized and lucrative sinecures used to ensure the loyalty of subordinates. Even lighthouses were originally provided by private concessions and funded by fees collected from ship owners.

Then, in the middle of the nineteenth century, everything changed. Suddenly new technologies started to make life predictable. When technology made it possible for ships to navigate further away from the coast, lighthouses could be transformed into public goods funded by the state. Tax dollars and state regulation could be employed to even out the problem of free-ridership. Steam power transformed economic productivity by providing a reliable source of energy. Canning made it possible to create safer and more reliable food supplies. Improved roads changed how people did business and created the potential for bigger capitalist marketplaces. Innovations in weights and measures gradually led to the replacement of an "infinite perplexity of measures" that differed by country, province, and even town with standard, regularized measures based on the swing of the pendulum or the tick of the clock or the fixed units of the metric system.[24] Gone were the hundreds of customary units of measure employed in eighteenth century France, crisscrossed as it was by a patchwork landscape of seigneurial authority. Even the foot (*pied*) varied from place to place. This metrical hodgepodge at once reinforced and reflected localized control, which entitled different aristocrats to impose their own system of measurement on each city or parish.[25]

From the mid-nineteenth to the early twentieth century, the industrial revolution precipitated an accompanying *institutional* revolution in what sociologist Arthur Stinchcombe calls social technology—the institutions that evolve in response to the broader scientific and technological reality. With the emergence of new technologies to measure and regulate time, energy, light, food, and other vital services, institutions designed around aristocratic patronage and private monopolies no longer made as much sense.

Instead, bureaucracy emerged as a type of centralized organizational structure with a rational division of labor. In this model, authority was based on administrative rules rather than on personal allegiance or social custom. Engineers, scientists, planners, and other professionals—what Max Weber called the "personally detached and strictly objective expert"—were trained to measure and manage. In a newly industrialized world full of new tools for controlling what was previously immeasurable and unpredictable, the delegation of authority to university-trained professionals offered a better, more efficient way to bring expertise to governance on a large scale than did the aristocratic or political patronage of earlier times.

Bureaucrats and professional officeholders were enlisted not only to instill efficiency and specialization in government but also to rein in the excesses of Gilded Age millionaires produced by the new industrial age.[26] Several independent commissions with professional bureaucrats were created to oversee railroads, substantially increasing the numbers of professionally employed civil servants.[27] Professional civil servants—many of whom might have been the children of conservative and wealthy farmers or tradesmen—abandoned their fathers' laissez-faire sentiment in favor of enthusiasm for stronger, more highly regulatory government that could curb the excesses of new corporations through professionalized administration. "If the philosophy and the spirit were new," writes Richard Hofstadter about this confluence of interests between different social classes in the Progressive Era, "the social type and the social grievance were much the same."[28]

Bureaucracy required predictability and valued continuity, which in turn depended on the new technologies of control, including common methods of measuring. These impersonal tools and rules to measure productivity were developed by a new middle class of modernizers, radically changing how society was perceived and governed. In addition, bureaucracy required officials with expert training, prompting the development of credentials and rule-based education. Both trends underpinned the emergent belief that public authority demands accredited specialists using standardized methods and measures.

The Standardization of Methods and Measurement

Enlightened aristocrats like Voltaire argued for the centralization and standardization of measurement to help advance France's position scientifically and economically. Initially, because the science of measurement was so closely tied to political institutions, this was an almost unfathomable prospect (or so said Diderot and D'Alembert in the entry on "weights" in the *Encyclopédie*). It was not until the French Revolution that Condorcet and his allies were finally successful in using political reform to push through the legal reform of weights and measures. The road to adoption was slow in practice but was aided, over time, by the tight fit between the needs of centralized, bureaucratic political institutions and the universal systems of scientific measurement.[29]

Common measures such as the meter were expected to accelerate trade and economic growth. The erosion of local customs and their replacement by national and international standards strengthened a sense of national citizenship and reinforced the power of the new nation-state (and later the Napoleonic Empire) in lieu of local feudal power. In fact, the adoption of the metric system is almost a proxy for the rise of the nation-state. Italy adopted the meter in 1861 and had adjusted its own local measures to be percentages of the meter. Spain went metric in 1869, Germany in 1872. Britain had its own system with the bushel, peck, rod, perch, and pennyweight, but eventually succumbed in the late 1960s. The metrification of Europe and, in turn, the ability to more effectively measure social and economic life were both caused by and the cause of greater political centralization. Much of early modern European statecraft seemed similarly devoted to rationalizing and standardizing life into more administratively convenient forms.[30]

In the United States, regularizing weights and measures was also a key development in the political centralization of the country and the rise of a professional bureaucratic class. In 1819, the House of Representatives, responding to the adoption of the metric system in France, asked Secretary of State John Quincy Adams to recommend a system of measurement for the new republic. In 1821 he recommended retaining the British—and rejecting the metric—system because of its long history and its association with the human body. "Of the origin of their weights and measures, the historical traces are faint and indistinct: but they have had, from time immemorial, the pound, ounce, foot, inch and mile, derived from the Romans, and through them from the Greeks, and the yard . . . a measure of Saxon origin, derived, like those of the Hebrews and the Greeks, from the human body."[31] Unifying measurement would eventually be essential to shoring up the authority of a new class of civil servants.

During this same period of rising industrialization and growth, mechanized clocks transformed the ability to measure time. Rationalizing public time was a much-talked-about development. As early as 1905, the United States Navy had sent time signals by wireless from Washington. The Eiffel Tower transmitted Paris time in 1910 even before it was legally declared the time of France.[32] Einstein was famously obsessed with the challenge of how to set a consistent time at Swiss railway stations.[33] Mechanized timekeeping and the ability to synchronize time between jurisdictions made possible the introduction of timekeeping for workers. Innovations like wireless telegraphy, the telephone, and the railroad all depended on and enabled accurate time. The recording of work time had earlier origins, but electrification and mechanization made it ubiquitous. Life became measured and ordered; work, appointments, and pleasure could now be organized, scheduled, and coordinated.

The newfound ability to manage time was reflected in the literature of the day, in which time featured centrally as a literary leitmotif. Many authors played with concepts of linear, stable time. Think of Oscar Wilde's narrator in the *Picture of Dorian Gray,* for example, or Proust's in *Remembrance of Things Past,* who experience extended time digressions while for other characters only a few minutes lapse. Kafka toyed with time in *The Trial,* in which the protagonist is never quite sure at what time he is supposed to be where, and the white rabbit of *Alice in Wonderland* is always late. The *Time Machine,* of course, was another work of this era. The irregularity of time for these literary figures arose from a broader cultural obsession with punctuality and a precision of timekeeping that had previously not been possible.[34]

Electrification enabled both accurate timekeeping and consistent illumination. Electric lights made it possible not only to ensure a reliable source of light, but also to blur the lines between day and night.[35] These advancements transformed the lives of city dwellers and the design of cities and towns, which could hold evening activities like meetings and movies. Electricity made people less susceptible to all manner of natural uncertainties and vagaries. They could regulate their environment for the first time, and society placed a "high premium" on regulatory stability and anything that would make relationships predictable.[36] Life became more orderly and capable of being ordered.

A whole host of innovations made the food supply safer, more plentiful, and more reliable. In the pre-industrial era before rail transport, most food was produced and consumed locally. Even with the railroad, transporting perishables such as slaughtered meat was not possible. Live cows had to be moved from the Midwest to the eastern cities for slaughter, which gave rise to a whole host of unpleasant urban conditions—and memorable novels describing them. Railroads invested heavily in the apparatus for loading and transporting live animals for slaughter and initially resisted innovations that would have made it possible to transport dead, dressed livestock. However, advances like the refrigerator car, coupled with the mechanization of the slaughterhouse, placement of abattoirs on the rails at the docks, and the creation of the "disassembly line" for rapidly cleaning a carcass while the object was in motion, suddenly made it possible to produce beef rapidly without any loss of quality.[37]

With the mechanization of tin can production came canned meat. Refrigerated railroad cars, which had previously transported meat carcasses packed in ice, were used to carry fruits and vegetables instead of meat, since meat could now more cheaply be transported in cans. The food supply became both more predictable as well as cheaper and more varied. Technology could now bring the farm to the consumer rather than the other way around, unlocking the opportunity for consumers to discover nutrition and opening up a whole new arena of regulatory activity.

This broad trend toward mechanization and industrialization—and the greater predictability of the conditions of production—led to the expansion of markets and economic growth. The adoption of new tools for measurement and new avenues for trade precipitated new kinds of economic specialization. This progressive differentiation of occupational functions became the basis for the rise of professions. Talcott Parsons described the professionalization of life in the nineteenth century as "the most important change that has occurred in the occupational system of modern society."[38] But industrial technology alone did not produce the inexorable wave of professionalization. It had help from several complementary historical trends: scientific positivism, the rise of the university, and economic demand.

During this period, advances in the natural sciences accelerated a paradigm shift toward rationalism and the application of scientific methods to the social sciences, including economic sciences and organizational psychology. Daniel Boorstin captured this Weberian demystification well when he wrote: "The old tricks of the miracle maker, the witch and the magician became commonplace. Foods were preserved out of season, water poured from bottomless indoor containers, men flew up into space and landed out of the sky, past events were conjured up again, the living images and resounding voices of the dead were made audible."[39] The growing ability to control the world around us eliminated some of its older magic.

What replaced the magic—and the unpredictability of life—were scientific positivism and the dawn of a new age of certainty and predictability. The new tools for measuring, standardizing, and regulating everything from energy to time to food production did much to level the institutions of aristocratic hierarchy and give rise to the professional, middle-class civil service trained by modern universities. Oliver Wendell Homes Sr. commented in 1860 that the two paradigmatic words of the day were *law* and *average*. "Statistics have tabulated everything—population growth, wealth, crime, disease. We have shaded maps showing the geographical distribution of larceny and suicide. Analysis and classification have been at work upon all tangible and visible objects."[40]

Like their natural science counterparts, fields such as economics, politics, and sociology aspired to graduate into real sciences and established unique disciplines with associated faculties.[41] This strengthened faith in a science of leadership and the need for an integrated system of administrative education, training, and professional development.[42] In turn, the technical specialization demanded by industrial age processes drove a belief in the need for a professional class of full-time public managers.

Frederick Taylor, who espoused the application of scientific principles and engineering methods to the improvement of work processes, is credited with inventing management science for the private sector. An engineer by training, Taylor sought to make manufacturing more efficient through the painstaking design of workplace practices. In his vision, administrators were essential for managing employees and assigning the right person to the right job based on competence and disposition. This view was not particularly flattering to workers, who were, according to Taylor not well suited to understanding "the real science" of doing this class of work. Instead, he espoused the need for elite managers.[43]

The impact of technology on institutions was characterized by two opposing trends. On the one hand, industrialization and professionalization created new institutions that were less aristocratic and more open to new middle-class entrants, including women. (Emperor Franz Joseph of the Austro-Hungarian Empire eschewed the use of typewriters.[44] Typewriters are credited with opening up the positions in public management in the United States to women, who were hired in great number into government in order to use them.)[45] The rise of professions is, in part, a story of democratization of "knowledge in triumphant practice." It is the story, writes Andrew Abbott, of "Pasteur and Osier and Schweitzer, a thread that ties the lawyer in a country village to the justice on the Supreme Court bench." It is about the evolution toward a great social mobility and openness on the basis of knowledge, merit, and achievement.

On the other hand, professionalization closed off participation to many by larding the system with rules, procedures, and expert jargon. Professional credentialing led to the exclusion of the amateur and his replacement by an authoritative, professional class. This view is consistent with the conception of professions as fundamentally about market control—a "chronicle of monopoly and malfeasance, of unequal justice administered by servants of power, of Rockefeller medicine men."[46] Professional communities did devolve into country clubs and fraternal lodges. The middle-class professions represented a democratic advance over aristocratic patronage, but they also ended up being exclusive and exclusionary in a different way. "Unchecked, the republic of letters becomes a republic of pals."[47] Barbara Ehrenreich caustically describes professionals as arrogant, self-serving careerists who purchase their status by discrediting everybody else as "amateurs."[48]

Combined with civil service exams and rigorous training in the new sciences of managing society, professionalism was believed to be the antidote to the corruption and venality of earlier government service. Professional bureaucracy was cheaper and better than patronage and privateering for maintaining social order. From the eighteenth well into the nineteenth century, judges charged transaction fees for cases they presided over. Naval officers were awarded a percentage of the value of the ships they captured. The fixed salaries for public servants we now take for granted, in fact, only slowly came to replace fees paid for public services and bounties awarded as compensation for a service rendered, such as capturing a prisoner.

Why depend on a loyalty system, which afforded plenty of opportunity for leakage and free-riding, when it was now possible to measure and manage time and space scientifically? Why rely on bureaucrats to negotiate fees with members of the public or accept bounties for performing a service for one social group at the expense of another when it was now possible to set and regulate salaries from above?[49] Technically trained professionals were thought able to decrease the costs per administrative decision well below the cost of decisions made by amateurs.[50] It now seemed clear that managers with fixed salaries and technical expertise were needed to run complex societies, and that the classical idea of amateur governance was obsolete.[51]

As the industrial revolution broke down the dominance of aristocracies, replacing them with a professional class of technocrats trained in the new tools of measurement, social life became thoroughly professionalized. Merit selection, specialized education, and training became the norm.[52] Salaries began to replace fees. The concept of merit-based civil service that emerged at the end of the nineteenth century embraced principles of competence, stability, and equality of opportunity.[53] New professional titles, honorifics, and professional associations mushroomed. In the 1890s, for example, the titles tree surgeon, sanitation engineer, beauty doctor, and mortician first appeared. Uroscoper evolved into urologist, and bone-setter into orthopedist. Crafts evolved into professions.[54] Morticians and dieticians, two position labels inspired by the word physician, saw their subjects, mortuary science and nutrition science, added to the curriculum of accredited colleges.[55] Morticians, often referred to as "doctors of grief,"[56] proposed that the newly formed National Funeral Directors' Association examine and license all members as professionals. They issued their own code of ethics. Professional organizations proliferated. In England, ten of the thirteen contemporary professions measured by one scholar acquired an association of national scope between 1825 and 1880. In the United States, eleven of the same thirteen were similarly organized into national associations between 1840 and 1887.[57]

Lawyers established their first national professional association in 1878; librarians in 1876; social workers in 1874. The American Social Science Association was founded in 1865, and the establishment of new curricula, journals, and specialized associations rapidly ensued. The professional social scientist emerged as the paramount and exclusive interpreter of man and society. Social science greats, including Weber, Durkheim, Pareto, and Tönnies, appeared on the scene not long thereafter. Virtually every medical subspecialty, from surgeon to neurologist to pediatrician, started its own association. So did scientists, chemists, and biologists. By 1870, Americans had committed themselves to a culture of professionalism. This was how they were going to govern large, complex institutions and territories. It was also how they were going to fight corruption, crony capitalism, and big city machine politics.

The University as Professional Training Ground

The professionalization of public life coincided with the rise of elite universities to train these new mandarins—a development that, in turn, reinforced the rise of scientific positivism. Formal training by those with professional standing emerged to teach those wishing to enter specialized disciplines. Instead of apprenticeship within the labor market and on the job, professional training for work became the norm—and the province of formal institutions devoted to organizing the knowledge within the profession's emerging discipline. Much of the perceived legitimacy of a profession's monopoly over a discipline stemmed from its claim to sort and order knowledge according to rational, scientific principles. Once accepted, such legitimacy reinforced the movement toward disciplinary specialization, as did placing the training for professionals in universities, which added the imprimatur of third-party certification as a prestigious measure of expertise. As universities established formal training programs to support and accredit new professions, they also solidified the related scholarship. Once the pastime of serious and wealthy amateurs like Charles Darwin, scholarship itself became a paid and accredited profession.[58] Previously called a natural philosopher, one who pursued science came to be called a scientist for the first time only in the 1830s. Professors formed their own professional guild in 1915: the American Association of University Professors.

All of these developments were consistent with a post-Enlightenment thirst for rationalism dressed up in scientific forms. This suited the general optimism and faith in progress that mark the later nineteenth century. In that environment, scientific rationalism appeared to be the perfect instrument to bring about the continuous improvement of society.[59] The seeds were already there. Indeed, the determination to base decisions on rational thought—that is, on reason rooted in scientific principles and supported by empirical measurement, rather than on the impressions of unreliable senses—had already begun with Descartes and the *Philosophes* in the sixteenth and seventeenth centuries. Cartesian thought made popular the notion that there were fixed laws of nature and mathematics distinct from qualitative reflections and that those laws could be proven through scientific experimentation.

The classic German terms *Wissenschaft* (science)—and its derivatives, *Wissenschaftler* (scientist) and *wissenschaftlich* (scientific)—originated in the nineteenth century. Professionals were men of science, whether university-trained as doctors, pharmacists, dentists, lawyers, or political scientists. The ideal professional, wrote Charles Eliot, Harvard's president in the late nineteenth century, was "neither learned nor practical but scientific."[60] They were now to be trained in deductive approaches to knowledge (what the Greeks called *techne*), not in older, craft-based approaches to their vocations or in practical knowledge *(metis)*. Formal curricula began to organize knowledge as a logical set of ordered principles supported by empirical observation. These professionals did not all wear white coats, but they were all scientists.

The universities that trained them were thoroughly middle-class institutions. For centuries, aristocrats had advanced as a result of purported superior abilities that were the result of heredity.[61] Universities expanded the pathway to advancement based on merit and training. Performance on examinations, not the social connections that led to apprenticeship opportunities, was now what counted. Linked intimately with the emerging professions, the development of higher education in America supported—indeed, made possible—a "social faith in merit, competence, discipline, and control."[62]

By 1870 there were more institutions awarding bachelor's degrees in the United States, as well as more medical schools and law schools, than in all of Europe. In the final third of the nineteenth century, the number of faculty members doubled in the first decade and then rose more than 300 percent between 1870 and 1900. In 1905, Andrew Carnegie established the first pension plan for university professors in secular institutions. This precipitated both further growth of universities and the conversion of many religiously affiliated institutions into secular ones. Where there had previously been only a small number of denominational colleges, there were now modern universities with libraries, laboratories, and large endowments.[63]

The number of undergraduate students jumped from 52,000 in 1870 to 355,000 in 1910. Graduate student enrollment multiplied even faster. In 1871, there were fewer than a few hundred graduate students in the United States. By 1910 that number had jumped to ten thousand.[64] Between 1868 and 1891, the wealthy philanthropists who founded Cornell, Johns Hopkins, Vanderbilt, Clark, Chicago, and Stanford donated more than $140 million to higher education. The

Morrill Land Grant Acts, which created the land grant universities, injected tens of millions more into the expansion of universities.[65] More than seventy colleges and universities, many of them in the Midwest, resulted from this legislation, which aimed to "teach such branches of learning as are related to agriculture and the mechanic arts, in such manner as the legislatures of the States may respectively prescribe, in order to promote the liberal and practical education of the industrial classes in the several pursuits and professions in life."[66]

In this climate, universities unsurprisingly began to transform the study of law, previously a vocational pursuit, into a field of scientific endeavor. At Harvard, Dean Christopher Columbus Langdell, who introduced the case study curriculum still dominant in law schools today, led the charge to convert American law from a craft taught to apprentices in law offices to a subject worthy of study at an elite university. Langdell developed the "casebook," in which students would first read the raw case material and then adduce principles from it. The goal was to train these new professionals in rigorous methods for classifying and sorting cases, in much the same way that a natural scientist works in the lab cataloging natural specimens. Law study, not law practice, was to be the way to train lawyers. A university was central to this task because it possessed a library. A law library is "the object of our greatest and most constant solicitude," wrote Langdell. The "library is the proper workshop of professors and students alike . . . it is to us all that the laboratories of the university are to the chemists and physicists, the museum of natural history to the zoologists, the botanical garden to the botanists."[67] The law was a science; the library its laboratory and repository; and its keepers professionals, whose decisions about the newly emerging canon of administrative law were absolute and not subject to judicial review.

The law was not unique in this regard. Universities co-opted myriad new fields. Harvard started the first school of dentistry within a university (as opposed to a vocational training college) in 1867 to expand the medical and scientific knowledge of what had previously been tradesmen. The Wharton School of Finance and Economy at the University of Pennsylvania was founded in 1881, a prelude to the subsequent declaration that business was a profession and management a science. Economics, although it emerged earlier, grew ninefold at the end of the nineteenth century.[68] In the 1880s, Columbia established its department of "political science" (the term had not been used before then) with the aim of "prepar[ing] young men for all branches of public service."[69] In 1886, the *Political Science Quarterly* began publication to cover the news of the "political science profession."[70] Chicago launched political science in 1900 and hired Charles Merriam as its first professor. Merriam, along with Woodrow Wilson, is largely credited with generating interest in public administration as a profession in its own right.

Beginning in the 1930s, schools of public administration emerged to combine the practical application of political science to management studies. The terms "public management" and "policy science" *(Staatswissenschaft)* come into use around the same time. Professional organizations, such as the Association for Public Policy Analysis and Management, the National Association of Schools of Public Administration, and the National Academy of Public Administration, appeared on the scene a few decades later, further solidifying a field that has, since its inception, drawn its power, authority, and independence from self-created laws and rules. Group after group sought to increase the status of its guild, plant its roots in universities, craft stringent requirements for admission, discover a "scientific" basis for its work, and set up firm exclusionary boundaries.

Given their new prominence, social scientists had to justify their place in the new professional world by demonstrating that their sciences could, in practice, achieve some level of truth. Therefore they advocated strongly for scientific objectivism as a disciplinary norm. Especially after the rise of behavioral social sciences like psychology, adherents sought to strengthen the professional position of these new disciplines by asserting that their legitimacy was rooted in observable, measurable phenomena.[71] They wanted empirical studies of human behavior, including human political behavior. "The experience of World War I accelerated the movement toward a new objectivism," writes intellectual historian Edward Purcell Jr.; "many social scientists had served in the government during the war, helping to bring centralized control and orderly planning to the management of the nation's effort."[72]

This new scientism, especially the professionalization of psychology and psychiatry and the popularity of Freudianism, implicitly challenged some key underlying assumptions of democratic theory: namely, the rationality of human nature and the practical possibility of a government of or by the people. For many, the experience of World War I was reason enough to lose faith in the rationality of man. Between Freud's popularization of the psychosexual and Le Bon's theories of mass emotionalism and crowd behavior, it was no wonder that a generation of political scientists emerged who questioned the democratic ideals of an earlier, simpler age.

Man's demonstrated inability to act rationally, reasonably, and wisely further reinforced a view that popular government was dangerous. The introduction of psychological approaches into the study of governance led, in some cases, to racist, antidemocratic, and even authoritarian theory, which heaped scorn on the intellectual inferiority of the masses—or at least of certain classes, races, genders, or types of people. Those who did not reject democracy out of hand believed that an administrative and scientific elite would be able to direct popular government along "rational and objective lines." Their belief that an administrative elite would be needed to cope with the increased complexities of industrial society did not celebrate but, rather, doubted what common citizens had to offer. In the face of the irrational, Purcell explains, social scientists came to view the use of experts as a matter of necessity, not convenience or wisdom.[73]

The Demise of the Amateur in Public Life

With America's growth in territory and population, not to mention its brutal tempering in the furnace of the Civil War, the need for a larger national government emerged in the nineteenth century. During the first half of that century, the population of the United

States doubled as the country expanded gradually westward, and federal government expenditures quadrupled and then more than doubled again between 1870 and 1910. The multiplication and complexity of laws involving new powers, new restraints, and new duties demanded an increase in the federal service. "The great increase of public records and documents . . . and the frequent calls of Congress imposing a necessity for researches, which comprehend the history and transactions of the department from it first organization, all contribute to render the duties of every officer and every clerk more difficult, complicated and laborious."[74] Initially, the civil service was small. A secretary of one of the original four Cabinet departments— State, War, Treasury, and Navy—typically had a staff of a dozen clerks and a few messengers, a limited field service, a small library, and no formal recordkeeping system. The federal government in 1800 comprised a total of three thousand employees; by 1860 their number had reached fifty thousand.

The wide enfranchisement of adult male citizens—and the increase in the middle class, which grew eightfold—fueled political activity, much of it increasingly partisan.[75] "No sooner do you set foot on American ground," wrote Tocqueville, "than you are stunned by a kind of tumult . . . almost the only pleasure which an American knows is to take a part in the government, and to discuss its measures."[76] The realization that success in elections would depend upon support from state and local party machines led to a politicizing of the public service. The ability to create and offer public sector jobs in customhouses and post offices to middle-class families helped to shore up political support. Unsurprisingly, family members of elected officials and political hacks began to populate the rolls of the public service.

Max Weber derided such a system: "Where 300,000 or 400,000 of these party men whose only demonstrable qualification for office was the fact that they had served their party well could not of course exist without enormous evils—unparalleled waste and corruption— which could only be sustained by a country with, as yet, unlimited economic opportunities."[77] Some sections of the public service retained the qualities of permanent tenure, expertness, and professionalism that make a career public servant. But political pressures meant the introduction of rotating political appointments that gradually came to dominate the civil service. The spirit and quality of the public service deteriorated markedly during the Jacksonian era, which was dominated by a "spoils system" of government by which presidents and their political parties granted their supporters the positions of governing.[78]

As part of the process of growing the civil service to increase the number of patronage-based positions, social functions previously undertaken by social service organizations were taken over by government. The earlier quality and strength of the civil service had been the result of public servants coming from a "class of gentleman who were men of integrity," who from 1789 to 1829 had accepted the duty of government. George Washington had emphasized appointing those local notables with fitness of character that, because they had the respect of their fellowmen, would both execute their administrative duties with integrity and competence and also engender loyalty to the new country. Washington himself eschewed payment for his own public service. By mid-century, however, men of sensitive and elevated character were being driven out of public service by politics and patronage.[79]

German sociologist Robert Michels has pointed out that pressure on organizations to create exclusive jobs and hire people to fill them leads inevitably to hierarchy. Even before the Jacksonian era, exploitation was common. In his landmark study of the salary revolution in American government, Nicholas Parrillo painstakingly details the varied forms of negotiated payments and bounties by which civil servants were paid before salarization became widespread in the late nineteenth century.[80]

Public service brought new middle-class entrants status and wealth—benefits that they inevitably moved to stabilize by professionalizing and thus shutting the door behind them. Increasing the number of full-time political professionals contributed to the perception that they possessed superior knowledge—an active interplay between economic need and the demand for professionalization. The lure of a public pension created additional pressure to develop more public sector jobs, forcing the bureaucracy in its own interest to "open the sluices of its bureaucratic canals in order to admit thousands of new postulants and thus to transform these from dangerous adversaries into zealous defenders and partisans."[81] With the rising size and, hence, political importance of an expanded middle class, new, secure sources of employment were needed. After the War of 1812, the major executive departments of Post Office, Treasury, and War added offices and functions. The trend continued. The dislocations caused by industrialization, and later the Civil and the First World Wars, also pushed these professional bureaucracies to grow.

Until 1829, there had been a single civil service system with no formal method of selecting its employees. After 1829, it bifurcated into the system we know today of political (rotating) appointments and career (tenured) public servants, but the distinction was not rigid, and patronage was increasingly the norm. Even petty clerks were rotated in an effort to dole out more jobs to party supporters. In the changeover from President Cleveland to Harrison, for example, more than 100,000 Americans rotated in and out of Post Office jobs.[82] The party patronage rotation system helped to preserve government against an entrenched bureaucracy, but it also led to a public service that was often "ignorant, clumsy, expensive, partisan and, at times, corrupt."[83]

Public service created the spoils of new, middle-class jobs for thousands of people, which was important in the face of economic costs that increased 35 percent between 1870 and 1910. The size of the national civil service quadrupled. Nonetheless, before the rise of a truly professional public service, the broad perception of the Andrew Johnson and Ulysses S. Grant eras was one of venality and scandal. A steady deterioration of efficiency and ethics ran hand-in-hand with a steady expansion and democratization of access to public sector jobs. Counterintuitively, the democratization of officeholding through the spoils system paved the way for the bureaucratization of the civil service. By disassociating the office from the person holding it, a federal appointment came to be linked to its functions rather than to the officeholder.[84]

In reaction to this backdrop of corruption, low ethics, and politicization—what Theodore Lowi described as 99 percent patronage and subsidy politics—there began to emerge the ideal of a system of administration that was businesslike in character, rather than political.[85] The shooting of President James Garfield by a disaffected office seeker and the resulting outcry over the perceived debasement of public service accelerated the move toward a professional and exam-based system of appointments based on merit rather than connections. Even before the Pendleton Civil Service Reform Act of 1883, agencies had been moving toward more businesslike forms of organization, salarization, and promotion.

Professionalizing public service and public policy roles was a clear moral and competence-based reaction to the earlier system of patronage. Woodrow Wilson's push for "a civil service cultured and self-sufficient enough to act with sense and vigor, but intimately connected with the popular thought, by means of elections and constant public counsel, as to find arbitrariness or class spirit quite out of the question" captures the spirit of the age.[86] It was also a political reaction to the increasing difficulty of controlling a many-headed hydra of local, rotating public servants—who were negotiating their own fees—from the national level. As the civil service grew, and took on new roles and functions, the spoils system became ungovernable and delegitimizing. Soon even the political elites sought to curtail the power of local political machines.[87]

Just as universities took training of lawyers away from law practices and turned law, political science, and policy science into formal academic disciplines, they also gradually redefined public service as a position demanding the expertise of university-educated professionals, rather than high school-educated clerks or hapless relatives. In time, university faculty started to be appointed to important government posts. As Leonard White explains, "As the newer disciplines of economics and statistics, political science, and the agricultural sciences found a place in the institutions of higher learning, specialists in these fields became useful in practical affairs."[88]

By the end of the nineteenth century, nearly half the graduates of American colleges were entering "teaching, administration, and business."[89] As a result, no other country in the world enjoyed as robust a profession dedicated to the teaching and analysis of government as the United States. The "only safety for democracy," wrote historian of political science Bernard Crick, "[is to] apply scientific methods to the management of society." Citizenship was not a state of being or a cultural mindset, but a formal discipline to be taught and credentialed. Woodrow Wilson took this idea to its natural conclusion by advocating for training in public management, not just in political theory or jurisprudence. His 1887 *Political Science Quarterly* article was the first recognition of public administration as a field worthy of study.[90] Young men needed to know the practical side of government and "whom to do business with in Washington."[91] To rationalize government was to rationalize society.[92]

But these benefits came with a cost. Although intellectual fashions come and go in the academy, the training of professional social scientists, lawyers, and policymakers has remained relatively stagnant. The curriculum of a first-year student at Harvard Law School today is nearly identical to the curriculum in Langdell's time: civil procedure, criminal law, contracts, property, and torts. The method of study remains consistent. Modern social sciences offer essentially the same set of academic departments and disciplines that they have for nearly one hundred years: sociology, economics, anthropology, psychology, and political science.[93] They have not been hotbeds of innovation. They have become the chosen instrument by which our society now creates, credentials, and sustains its own elite class of mandarins.[94]

How Professionalism Subverts Participation

Limited innovation is not, however, the only cost. The history of governance is the history of what political anthropologist James Scott calls social legibility—the mechanisms for reading social conditions to the end of regulating them more effectively. The modern professional is trained in the name of scientific principles to measure, quantify, regulate, and bring order to a society previously unsuited to management by centralized institutions. As Scott explains in *Seeing Like a State,* his magisterial tour d'horizon of failed attempts by authoritarian states to impose grandiose visions of social ordering, the job of governing requires techniques to enhance understanding of complex and vast social and economic systems. This ability to classify, quantify, and predict goes hand in glove with the rise of the professional civil service and the solidification of the nation-state. Measurement, mapping, and census taking are all strategies that centralized governors use to know whom they govern and to simplify their understanding of conditions on the ground. Over time, the project of standardizing complex and diverse conditions—of mapping and measuring—however, can impose a false description of reality. If that happens, legitimacy and effectiveness are threatened.

In Scott's example, a forest manager in nineteenth-century Germany classified and counted trees in order to account for their commercial, not their eco-systemic or botanical, value. By measuring incompletely, the manager facilitates the short-term goal of taxing the production of wood but undermines the long-term goal of successful governance. Experience teaches that governing well over time requires a thoroughgoing but flexible understanding of the reality of on-the-ground conditions. Accordingly, Scott laments the unalloyed spread of bureaucratic logic.

Any political system in which experts dominate, wrote Socialist political theorist Harold Laski in the early part of the twentieth century, develops the "vices of bureaucracy" because the professionals will push "private nostrums in disregard of public wants."[95] In its drive

to simplify in order to count and measure, such professionalism—
back to our German forest—maximized economic return from the
wood but failed to take account of the true complexity—and therefore the myriad forms of value—inherent in an old-growth forest:

> A diverse, complex forest, however, with its many species of trees,
> its full complement of birds, insects, and mammals, is far more
> resilient—far more able to withstand and recover from such injuries—
> than pure stands. Its very diversity and complexity help to inoculate
> it against devastation: a windstorm that fells large, old trees of one spe-
> cies will typically spare large trees of other species as well as small
> trees of the same species; a blight or insect attack that threatens, say,
> oaks may leave lindens and hornbeams unscathed. Just as a merchant
> who, not knowing what conditions her ships will face at sea, sends out
> scores of vessels with different designs, weights, sails, and navigational
> aids stands a better chance of having much of her fleet make it to port,
> while a merchant who stakes everything on a single ship design and
> size runs a higher risk of losing everything, forest biodiversity acts like
> an insurance policy.[96]

Centralized states are highly vulnerable to this kind of blindness.
Their very professionalism fills them with what the economist and
sociologist Thorstein Veblen decried as "trained incapacity." By design,
professional ideology is blind to the rich and complex tapestry of
people's diverse capacity. Of course, bureaucratic organizations de-
rive their power, in part, from the support of coalitions that favor
their policies. The Post Office aggrandized its own power in the late
nineteenth century, even in the face of congressional opposition, because the support of Progressive moralists, agrarians, business interests, and immigrants, who distrusted private banks, provided the
agency with the latitude to develop its policies.[97] But the moral protection of such diverse interests gave this and other agencies *more*
autonomy and *more* power to close ranks around the vision of reality
that it held.

Bureaucracies, with their own rules, language, customs, and practices, use an artificial measuring stick to evaluate people's intelligence. Those who fail to speak the language of this high priesthood, who are not trained in their academies, and who do not conform to the cultural practices of their profession are deemed too ignorant to govern. Thus, as universities became the way to certify know-how, decrease the costs of finding such expertise, and increase trust in governing, they obscured the role of citizen experts—and put up barriers to keep them outside the doors behind which the real work of governance gets done. It brought regular competence and order but at the price of an exclusionary monopoly that replaced an aristocracy of social position with an aristocracy of esoteric knowledge. After all "a profession has many of the qualities of a self-perpetuating monopoly— in this case, a monopoly in authoritative opinion about the nature of man and society."[98]

Given the challenges our society faces today, however, we may well be at the tipping point where this cost becomes unacceptable. Professions build high walls. They develop specialized vocabularies, customs, and rituals that exclude others from the walled garden. They use their symbols of professional authority to widen the gulf with others. Today the real source of worry is less corruption and self-interest (with notable exceptions) than the inability to apply the full range of relevant, available expertise to our most pressing challenges. In his day, Max Weber could laugh off participatory approaches to governance as hopelessly inefficient. What was needed, instead, was a class of elite public managers—the best and brightest few good men—who could use the latest scientific tools to govern in the public interest. Today, however we very much need what citizens have—but what our professionalized institutions of government are precisely and venerably configured to prevent. We need to tap the wellspring of our deepest resource—the collective imagination and ingenuity of our democratic citizenry. But after a century of professionalism, our institutions are willfully blind to what our people are capable of and willfully ignorant of their eagerness to contribute.

3 The Limits of Democratic Theory

The private citizen today has come to feel rather like a deaf spectator in the back row, who ought to keep his mind on the mystery off there, but cannot quite manage to keep awake. . . . Rules and regulations continually, taxes annually and wars occasionally remind him that he is being swept along by great drifts of circumstance. Yet these public affairs are in no convincing way his affairs. They are for the most part invisible. They are managed, if they are managed at all, at distant centers, from behind the scenes, by unnamed powers. . . . He lives in a world which he cannot see, does not understand, and is unable to direct.

—Walter Lippmann, *The Phantom Public*, 1925

In 2012, the White House established an online petitions website, We the People.[1] Like its British counterpart, e-Petitions, the U.S. website allows members of the public to solicit signatures in support of a petition for government action. Petitions have been launched on such diverse topics as enacting federal gun control reforms, legalizing marijuana, supporting mandatory labeling of genetically engineered foods, and building a "Death Star," like that in the movie *Star Wars*. Initially, only 5,000 signatures were needed to prompt a response from the executive branch, but the popularity of the system—it is the only way to communicate with the White House directly other than by mail—led to the requirement being raised to 100,000 signatures in thirty days.

We the People is designed to reach large audiences and connect them to the White House. As a means of engagement, the petition website can sometimes enable individuals and interest groups to raise awareness of an issue among media or political elites, as was the case with unlocking cell phones in 2013. At that time, phone companies could prevent consumers from unlocking their mobile phones for use on a competitor's network without the carrier's permission, even after the consumer's contract had expired, because of the interpretation of a provision of the copyright law that prohibited circumventing devices to protect against piracy. A petition on the website demanded that the White House call upon the Librarian of Congress to issue an exemption to that provision, rescind the law, or at least to champion legislation to legalize cell phone unlocking. The petition garnered 114,000 signatures and drew significant attention to the far-reaching impact of a previously obscure and technical copyright policy.

Petitions can get a topic on the public agenda by opening a channel of communication other than lobbying or appealing to congressional officials.[2] Humorous petitions like the demand to construct a Death Star or deport the teenage heartthrob Justin Bieber back to Canada might, at most, force a press release by the White House. But even the most poignant petition has no real impact on policymaking. When the White House made the data from the website available in May 2013, We the People had hosted 200,000 petitions with 13 million signatures, yet only 162 had received a response—and none can be directly connected to a decision made, dollar spent, or action taken by the government.[3] A year and a half later, at the end of 2014, there were still only 165 responses from the White House. It is probably no wonder that use of the site dropped off precipitously. Few petitions reached the required signature threshold, and close to half of the petitions made requests for actions outside the scope of presidential powers.[4]

We the People does nothing to help already beleaguered policymakers identify innovative new ideas for responding to complex issues such as mental health, school safety, or gun control. It does not ask the public to contribute cogent solutions, quantify their costs and benefits, weigh alternatives, amass supporting data, describe their experience with similar policies in other jurisdictions, determine who will be affected, identify experts, or point out practical and political pitfalls. Even if the government were willing to build a Death Star, the website—really just a suggestion box—would offer no implementable guidance about how to do so. No one is invited to share relevant knowledge, experience, or expertise.[5]

At their core, efforts to solicit electronic petitions or encourage citizen questions or complaints via the web are based on the Platonic view that citizens possess neither the knowledge nor the competence needed for decisionmaking or constructive action. Plato famously argued in *The Republic* that democracy was flawed because ordinary men, in contrast to philosopher kings, do not have the capacity to make judgments about the common good. Similarly, In "Federalist 57," James Madison, like Rousseau before him, suggested that the people's virtues lie primarily in their capacity to choose for rulers "men who possess most wisdom to discern, and most virtue to pursue, the common good."[6] In this view, knowing one's own self-interest does not translate into the ability to dictate the public interest. Our leaders know best. They should be the ones to decide.

With the rise of the administrative state and the ideology of professionalism in the late nineteenth and early twentieth centuries, the day of Harold Laski's plain man has passed. "No criticism of democracy is more fashionable in our time than that which lays emphasis upon his incompetence," declared Laski. The plain man cannot plan a town or devise a drainage system, or decide upon the wisdom of compulsory vaccination without aid and knowledge at every turn from men who have specialised in those themes."[7] In other words, either we trust the professionals, or government will break down altogether.

This view predominates today. To the extent that the public does have a role in public administration, it is very small, mostly limited to legitimizing official decisions after they have been made, but disconnected from actual decisionmaking. Occasionally, citizens are asked to comment on pending regulations or serve on scientific advisory committees, but their role is primarily to talk, not to act or decide. These kinds of participation are to real engagement as "Kabuki theatre is to human passions," writes the former general counsel of the Environmental Protection Agency (EPA), E. Donald Elliott. They are "a highly stylized process for displaying in a formal way the essence of something which in real life takes place in other venues."[8]

It is a commonly held view that the average voter does not play a role in making policy or participate substantively in democratic affairs beyond the ballot box. The day-to-day running of the state is the domain of professionals, both in practice and as a matter of political theory. Numerous political theorists, from Jürgen Habermas to Claus Offe, have likewise argued that the "possibilities and benefits of participation are now seriously limited by technological and social complexity."[9] Robert Dahl goes so far as to call any suggestion that ordinary people have the capacity for deeper participation in policy-making on complex topics "extravagant."[10]

Of course, not all political theory is equally pessimistic. In contrast to the assumption that governance is the domain of elites, progressive scholars like Benjamin Barber and communitarians like Amitai Etzioni and Harry Boyte argue for an expanded role for citizens in self-government. Deliberative democrats such as James Fishkin, John Gastil, James Bohman, Amy Gutmann, and Dennis Thompson are also far less skeptical about the potential for citizen participation. But in general, much contemporary political theory— even participatory democratic theory—fails to identify concrete, institutional opportunities for genuinely substantive participation in *formal* decisionmaking processes.[11]

Hollie Rousson-Gilman's work on the micro-politics of participatory budgeting is unique in opening our eyes to the potential for citizens to be "architects of their own involvement" in hard and real decisions about how cities should spend money.[12] Archon Fung's corpus of work on citizen engagement includes examples of reinventing urban democracy in Chicago, where nonprofessionals are making a real difference in the administration of public life.[13] Yet most political thought is dismissive of citizen competence. This dominant view sees substantive citizen engagement, as distinct from providing oversight or volunteering, as being at odds with good governance.

The litany of complaints is as long as it is familiar. Citizen participation does not lead to effective decisionmaking or problem solving. It does not work because people lack time, education, and motivation to participate in ways that are helpful. Furthermore, such engagement is unnecessary because interest groups amalgamate the views of citizens more efficiently and productively. Direct participation only adds "noise" to the signal.

Others dismiss citizen participation as a sham, not because of any incapacity on the part of the public per se, but because ultimately those in power decide, often in secret, and often in ways that are driven by party politics. In any case, unless government officials are willing to admit that they need outside help formulating solutions, engagement will never work. A related complaint is that reform efforts need to tackle the problems of corruption, partisan politics, and power rather than just promote citizen engagement. There is, after all, no shortage of information today, just an excess of politics. Opening the floodgates to more participation could even increase corruption, bias, and regulatory capture by exposing disinterested officials to the sway of public influence.

In addition to these complaints, some dismiss citizen engagement as a form of techno-utopianism. They say there are no models for engaging people in making policy in large numbers. The public can serve only to ratify decisions after the fact. To suggest otherwise reflects a naïve belief in the potential of technology.[14] To the extent that we can organize interaction at a distance and, as one think tank predicted for 2015, democracy will "make itself at home online," public engagement in the administration of the state will add little.[15] It may help in tackling small or minor problems—such as the odd challenge or petition—but not important challenges.

As Chapter 2 noted, in the late nineteenth century, educated professionals began to view themselves as far better experts in the science of government than gentlemen public servants. People with degrees, accreditation, and legitimacy were needed to run things, and a professional civil service trained in modern universities by their new social science departments provided the members of this new, highly professionalized governing class. A natural corollary of the ascendance of educated professionals within government was the exclusion of amateurs from formal governing. We may bemoan the lack of trust citizens have in their government today, but no survey measures how little any government actually trusts citizens.[16] Our dominant political theory has long been deeply distrustful of citizens and their abilities, and so provides little support for redesigning institutions to take advantage of their expertise.

This chapter examines the prevailing critiques of citizen illiteracy, pathology, and inefficiency in contemporary political theory, which further buttress the preservation of closed-door institutions and the exclusion of citizens from governance. Despite clear evidence that professionals do not always "know better," are not always impervious to outside influences, and do not always possess all the best ideas, these judgments about the public persist because of, not in spite of, democratic political thought.

Citizen Illiteracy

The question of citizen competence has a long history in liberal thought. Restrictions on the franchise and the exclusion of the broader population from day-to-day decisionmaking have often been justified by the claim that average citizens lack the requisite skills or intelligence.[17] People, in this view, do not possess the training, disposition, or ability that professionals do. They are chronically misinformed or underinformed, or congenitally incompetent and irrational.[18] This belief in citizen ignorance shaped John Stuart Mill's views on effectively limiting the franchise by granting more votes to better-educated voters. The historical roster of advocates for replacing democracy with epistocracy (rule by those who know more than others, such as experts) includes Plato, Aristotle, Freud, and even Jefferson, who believed in a "natural aristocracy" and who extolled the virtues of educated elites and expressed doubt about the capacity of the public at large to engage in governance.[19] Although the Founding Fathers believed in popular elections, they vigorously argued in the *Federalist Papers* that only educated men of a certain class should or could run affairs of state. "The Federalists," writes the famous historian of government Leonard D. White, accepted the philosophy of government for the people by a "superior part of mankind," but not government by the people.[20]

More recent scholars in the rational-choice tradition have focused on political ignorance. Bryan Caplan and Ilya Somin, both law professors at George Mason University, following in the tradition of Joseph Schumpeter, Morris Fiorina, and Anthony Downs in his 1957 classic *Economic Theory of Democracy,* see the basic act of voting as irrational and hence do not even arrive at more substantive questions of citizen engagement.[21] More recent works on voter rationality are replete with examples of people's ignorance about contemporary politics. In *Just How Stupid Are We?,* Rick Shenkman points out that 50 percent of Americans can name four characters from the Simpsons, but only 40 percent can name all three branches of government.

Somin reels off the following unsurprising statistics: Only 24 percent of Americans understand that cap-and-trade is an environmental policy; many think it refers to health care reform or regulating Wall Street. Only 15 percent know that David Cameron is prime minister of England. Only 28 percent know that John Roberts is chief justice of the United States. Only 73 percent know that Congress passed a health care reform bill in 2010. Examples of such ignorance make great press. Citing them allows some commentators a fairly safe way to suggest that ignorance is linked to groups with lower socioeconomic status or to racial and ethnic minorities. If voters are "worse than ignorant"—Caplan decorates his cover with pictures of sheep—it is entirely sensible for them to stay home on Election Day.[22]

In an effort to rescue democracy from ignorance—and defend the franchise—Somin contends that ignorance may be rational, but that it is also quite harmful. Holding government accountable requires that voters have some knowledge of how it works. He *wants* informed voting. He just does not think he can get it. His normative complaint is ideologically driven. Without an understanding of government, voters cannot agree to his prescriptions for privatizing it. Or, put another way, because people are so ignorant and cannot provide adequate oversight, it makes sense to have smaller government. Because we cannot police it, we should get rid of government and give people as few opportunities as possible to ruin things.

Somin also contends that, from an efficiency perspective, it is smart to be ignorant. When a single vote matters so little, political ignorance is economically rational and efficient. It makes no economic sense for citizens to devote time and effort to political participation or acquiring political knowledge.[23] Not only conservative thinkers but also progressive political theorists, such as Carole Pateman and advocates of "thicker" forms of political engagement—which demand more active participation from citizens than "thin" activities like petitions and voting—have written about the "infinitesimally small" impact of a single vote in an effort to make the case for more robust and impactful forms of citizen participation.[24]

Although the leap from voter rationality to privatization and decentralization is long and tenuous, there is an important current running through the literature on citizen competence, *both* progressive and libertarian: voter intelligence is measured by knowledge of political sport, not social problem solving. When theorists write about citizen ignorance, they are referring to the common inability to name a senator or identify the branch of government that declares war. Somin, for example, defines political knowledge as "awareness of factual matters related to politics and public policy. This includes knowledge of specific policy issues and leaders . . . broad structural elements of government . . . and the elements of competing political ideologies." The relevant knowledge to be measured is limited exclusively to politics and civics. In the same vein, you cannot report on the World Series if you cannot follow the game.

When these political scientists and economists complain of ignorance, they are limiting their analysis to a positive description of political facts, not a normative descriptive of cognitive ability. For instance, there are four major knowledge prerequisites for citizenship, according to Somin: understanding which problems are caused or can be alleviated by government policies; knowing the names of incumbent officeholders; knowing basic facts about an incumbent's term; and knowing whether the incumbent's policies were the best choices available.[25] In other words, political expertise is entirely self-referential and defined by understanding the "inside baseball" of current institutional arrangements. Somin and others who focus on voting behavior tend to neglect the broader range of human capacity and downplay its relevance.

Even scholars who do not conflate people's ignorance about the sport of politics with a broader condemnation of their skills and abilities neglect the possibility of a world in which nonprofessionals govern by bringing skills from "the real world" to solving public problems. The literature on citizen illiteracy takes voting for granted, questioning the capacity of citizens rather than seeing the anemic quality of the act of voting as the problem. Even deliberative democrats who focus on creating informed polls, and who have argued for establishing a national holiday dedicated to political debate, focus on educating citizens about politics as their aim, not being educated by citizens about problem solving. The measure of civic expertise is calibrated to government, but not to governing.

The fact that people might possess skills and know-how relevant to governing that could be used to solve problems, create public goods, or deliver services simply does not factor into consideration of civic intelligence. Their expertise, or experiences, with health care or transportation or education, or their skills in agriculture or science or ditch digging, are not part of the discussion because there is not even a glimmer of recognition that political participation in governing could extend beyond voting. Although Japanese philosopher Hiroki Azuma claims that information technology does not make democracy possible but, instead, that technology changes the very nature and meaning of democracy itself, this is not the common view. For most, democracy means "reacting to the doings of politicians with cheers and jeers."[26] Voting is the ne plus ultra of political expression. Seymour Martin Lipset, in his canonical *Political Man,* defines the hallmark of democracy as "a system of elections." The way people govern democratically is by choosing between candidates.[27] The general population constitutes the "clients and consumers" of the business of politics.[28] End of story.

Deliberative Pathology

Another school of thought discredits popular engagement not on the basis of *individual* irrationality but on the basis of the pathology of group dynamics.[29] According to these "group critics," we cannot organize participation because we cannot create workable and deliberative groups at any scale. Group dynamics are as likely to be poor as to be good. Once people get together, whether in real space or online, the dynamic can easily become pathological. Although group deliberation and discussion are celebrated as an inspiration in democratic theory, actually engaging individuals in groups creates pitfalls, including group homogenization and polarization, that must be avoided.[30]

Attention to group pathologies is the outgrowth of group research by social psychologists, which was first made popular in the 1940s and has since spread to communications departments and business schools.[31] In 1945, Kurt Lewin established the Research Center for Group Dynamics at MIT, devoted to conducting controlled experiments in group communication, in both laboratory and field settings.[32] Lewin's work was a reaction to fascism. Sigmund Freud and Elias Canetti, writing in the German-speaking world (and José Ortega y Gasset, writing in and about Spain) had pointed to the dangers of group association after the economic dislocation of World War I, the collapse of the Weimar Republic, and the rise of fascism. They observed that even otherwise benign individuals, when joined together under the sway of a charismatic leader, can become dangerous.

The center's researchers studied how the mission, goals, and structure of groups affected their functioning. They looked at authority relationships and leadership and assessed the entrainment, pacing, rhythm, and longevity of groups for clues about how to orga-nize them well. It also studied, often mathematically, the relationships within and across groups and networks. For example, one method for studying group behavior was to have participants sit in adjoining cu-bicles where they could not see each other but only pass written mes-sages. Then the researchers constrained who could pass notes to whom, thereby manipulating the pattern of ties and flows of infor-mation. This allowed them to study the impact of "various network configurations (e.g., circle, chain, y, wheel, and all-channel) on group functioning and performance."[33]

Emerging from this research is the extensive literature in cogni-tive social psychology and behavioral decision theory (about how people make judgments, decisions, and choices) that has revealed the challenges and dysfunctions of successful group dynamics.[34] This re-search has produced trenchant insights into how *some* groups func-tion or fail to. It has deepened knowledge about the fallibility of human decisionmaking, especially in teams, and the biases to which people are subject.[35] It has gathered empirical proof of the proclivity, for ex-ample, of people to ignore evidence that contradicts their precon-ceived notions.[36]

Although these insights have not been widely applied to collec-tive reasoning and group decisionmaking in *public policy* settings, much work has been done on the dysfunction and rancor of *political* discussion and debate. Research shows that statements made by people in political discussions—online and off—may reduce other people's willingness to express their own opinions, and that distortions of opinion may result from a fear of social isolation. Noelle-Neumann for example, has argued that media depictions of the normative "climate of opinion" have a silencing effect on those who hold mi-nority viewpoints; creating a spiral of silence that successively em-boldens the majority and enervates the minority. In the 1970s, Irving Janis famously described how interaction in groups can produce shifts of opinion to more extreme positions than most indi-viduals would actually prefer (groupthink). The same phenomenon occurs in web-based discussions where, for example, the argumentative climate of online discussion has been shown to change post-discussion opinion.[37]

Large-scale data sets from social media, such as Twitter, now make it possible to apply clustering algorithms to detect patterns of flow and influence. The research that has resulted offers continuing evidence of political polarization and self-segregation.[38] Conversations on Twitter create networks with identifiable contours as people reply to and mention one another in their tweets. As the Social Media Research Center explains in a joint study with the Pew Research Center, these conversational structures differ, depending on the subject and the people driving the conversation. Six structures are regularly observed: divided, unified, fragmented, clustered, and inward and outward hub-and-spoke. "If a topic is political, it is common to see two separate, polarized crowds take shape. They form two distinct discussion groups that mostly do not interact with each other."[39]

How to foster productive social interaction while avoiding untoward group influence has occupied a large place in writings on public opinion and democracy.[40] The challenges of information cascades, groupthink, and other organizational psychology problems of group dynamics darken the prospects for greater engagement. Sometimes critics focus on the design flaws of how groups are organized and run; sometimes they suggest that problems of group dynamics cannot be mitigated and so voice a deep-seated fear that bringing people together will inevitably be disastrous.

Using empirical experiments, sociologists and psychologists have linked crowding to anxiety, social disorder, and hostile behaviors.[41] They have uncovered compelling proof to shore up Madison and Le Bon's argument in earlier times about the unfortunate tendency of groups to become factions. This concern about how people behave in groups has led many to be suspicious—indeed, highly suspicious—of what modern citizens will do if allowed to associate and participate.[42] But once again, the idea that politics, which is intended to be a partisan and rancorous fight, is correlated with predictions about policymaking and problem solving is not borne out by empirical proof. "Bad behavior" in one context, especially in a lab setting and in the absence of participatory mechanisms in real institutions, does not necessarily translate into a vision of the future of the state.

Participatory Inefficiency

A third assumption about the limits of engagement focuses on inefficiency rather than irrationality or pathology. Complaints about voter irrationality often take it for granted that professionalism is the only alternative. Similarly, those who argue that professional organizations are more efficient assume that people are not capable of participating well, whether for reasons of cognitive incapacity or lack of time (or the tendencies of group dynamics). The two lines of thinking are deeply intertwined. But the efficiency argument is a distinct and important thread that dominates twentieth-century thought in both economics and sociology. It grows out of experience with an industrial economy dominated by large, hierarchical corporations that, in turn, helped to shape the government institutions designed to regulate them. In this vision, modern society is simply too complex for citizens to participate efficiently.

The assumption of professional hierarchy as the only efficient way to organize institutions of governance has shaped our limited conception of what the democratic public can do.[43] Hierarchy in and of itself is but is one way to organize the flow of information and control in a system. Although sociologists distinguish between professions, on the one hand, and bureaucracy (hierarchical workplaces where managerial authority drives worker behavior and advancement) on the other, the concepts often work closely together in practice.

Hierarchy is a ubiquitous mechanism for organizing distributed information flows in biological as well as sociological systems. Information flow theorist and Duke University engineering professor Adrian Bejan explains that design, whether of institutions or natural systems, is a "fundamental law of physics, not just of biology."[44]

Whether corporations, snowflakes, governments, or rivers, all systems pulse with movement, with flows into and out of the system, whether those flows are of heat, electricity, or information. To know why things look the way they do, he explains, first recognize what flows through them and then think of what shape and structure would best facilitate that flow. The relationship between flow and structural design is what he calls the "constructal law," which, he asserts, is as binding as any basic law of physics. It applies to all of nature, including social and political institutions. In this view, hierarchy is a logical way to manage information flows in a large, pluralistic society. Just as in biological systems, where hierarchy is an efficient, non-value-laden method of organizing flow, the scale of complex social systems makes hierarchy inevitably the right design for governance in large societies.

The view that the scale of modern society demands hierarchical forms of organization is consistent with the view of size—in this case small size—present in classical democratic thought. Aristotle said that democracy is the distance man can traverse in a day because the polis must be small enough to know one another's character and virtues.[45] Mill believed that "since all cannot, in a community exceeding a single small town, participate personally in any but some very minor portions of the public business, it follows that the ideal type of a perfect government must be representative."[46] Oxford anthropologist Robin Dunbar famously popularized this intuition when he put forward the argument—now known as Dunbar's number—that no person can maintain active social relationships with more than 150 people at a time. But how does that concept apply to democracy? How do we preserve the convivial bonds of social participation in large, complex societies?

To be clear, Bejan is an engineer, not a political theorist, and expresses no view on citizen engagement and participation, only on natural systems. However, hierarchy in social systems, too, has been the archetypical response to managing large-scale institutions over the past hundred years—perhaps most clearly seen in the rise of the centralized, hierarchical, vertically integrated corporation. Headquarters set goals, and specialized subunits carry them out. This is essential to efficient governance and social ordering. It is "no accident," writes Nobel prize–winning economist Oliver Williamson in some of the best cited works in the economics literature, "that hierarchy is ubiquitous within all organizations of any size . . . non-hierarchical modes are mainly of ephemeral duration." Williamson, who studies the relationship between markets and hierarchies as a means of organizing production, is one of many economists who have focused on the role of hierarchies to counter collective-action problems of free riding, in which one actor abuses the commons by consuming more than her fair share, or hold-ups, when a lack of mutual trust prevents otherwise rational collaboration and cooperation.[47] By bringing the activities of a system inside the boundaries of a traditional firm, transaction costs can be reduced.

Both sociology and economics have contributed to the view that hierarchy is unavoidable. Shortly before World War I, German sociologist Robert Michels wrote about the organization of left-wing socialist parties. Though they espoused egalitarian ideals, even these left-leaning parties tended not toward democratic arrangements but toward oligarchy, Michels pointed out. He concluded that oligarchy is an intrinsic part even of democracy. "It is organization which gives birth to the domination of the elected over the electrons, of the mandatories over the mandators, of the delegates over the delegators. Who says organization says oligarchy." There is a force of social and political gravity that makes bureaucracy the "sworn enemy of individual liberty, and of all bold initiative in matters of internal policy."[48]

To Michels—at the time an avowed socialist—the desire by left-wing as much as right-wing parties to amass power was not rooted in material class differences but in the very nature of organization itself: that the need for bureaucracies to govern large, complex societies inevitably leads to a centralization of power. Ironically, Michels saw charismatic leadership as the best escape from the iron law of oligarchy, and he became an ardent follower of Mussolini at the end of his life, believing that only a charismatic leader has the ability to break through an organization's inherent conservatism to excite the masses and do great things. In the end, Michels bet that the only real path to social change was through charismatic leadership, not democracy, since democracy ultimately had to lead to bureaucracy, which only stultified creativity. Though his conversion to fascism clearly put him on the wrong side of history, Michels is paradigmatic of many writers, including the American progressives, who thought it is simply not possible to govern on a large scale by any means other than hierarchical institutions—and certainly not by broad public participation. The processes of modern technology and machine industry—not to mention the complex tasks of civic life—make hierarchy necessary.[49]

Michels might indeed have been right at the time. Before the Internet, it was not possible to conduct the flow of information between government and citizen. Large-scale direct democracy was simply unrealistic. As Michels put it: "It is obvious that such a gigantic number of persons belonging to a unitary organization cannot do any practical work upon a system of direct discussion. The regular holding of deliberative assemblies of a thousand members encounters the gravest difficulties in respect of room and distance; while from the topographical point of view such an assembly would become altogether impossible if the members numbered ten thousand."[50]

President Woodrow Wilson, whom most consider the father of modern public management science in the United States, accepted without extensive analysis or critique the "advisability of hierarchy and the orderly arrangement of superior and subordinate positions, the unity of command, organizational discipline, 'scientific' principles and practices in financial and other forms of management, merit selection, professionalism, specialized education and training, the cultivation and functional deployment of technical expertise, and even the fostering of organizational *esprit de corps*."[51] Like other progressive leaders of his day, Wilson believed that modern industrial society had grown too complex for the common citizen and the average politician.

Even—or perhaps especially—progressive thinkers who embraced the value of educating citizens still took hierarchical management as a given and conflated hierarchy with professionalism. Progressives like Wilson and John Dewey propounded the view that more methodical, prudent, and disciplined hierarchies would lead to better organization of expertise in policymaking. For all their faith in hierarchy, in their progressive worldview there still existed an unbridgeable divide between experts, understood as professionals, and the public, who were anything but.[52]

More recently, however, writing about firms and hierarchies has exploded as a result of new thinking about the decentralized, networked forms of organization by scholars such as Yochai Benkler, Steve Weber, and Henry Chesbrough in their work on peer production, open source software, and open innovation consumer communities. As this work, which focuses on the economic value generated by user self-organized, bottom-up communities, suggests, the iron law of bureaucratic hierarchy is perhaps not completely unbending. In the cases they analyze, the hierarchical models of organization characteristic of industrial society are being displaced by the institutional arrangements made possible by a networked information society.

With the transition from centralized broadcast models of communication to an Internet ecosystem where everyone can be his own broadcaster and information producer (albeit often enabled by a commercial platform) on Medium or Twitter, new forms of decentralized individual action are emerging through nonmarket means, such as Wikipedia and OpenStreetMap, which allow people to participate occasionally on their own schedules. In part, these peer-to-peer systems are possible because they do not require hierarchical forms of organization to channel information flows. Some of these systems defy classical assumptions about the profit motive. But even for-profit platforms (think peer economy companies like the Uber transportation platform and the Airbnb platform for housing rentals by owners) organize themselves in a flatter structure. They rely on software to coordinate activity but without the need for bureaucratic hierarchy. In a networked information society, even a highly capitalistic one, rigid organizational forms are no longer always necessary for people to act, either by themselves or in concert.

Other, more "popular" recent works also attack hierarchical control in an effort to debunk it. Nassim Taleb's *Anti-Fragile,* for example, criticizes the role of bureaucratic hierarchy (and just about everything else) for enabling rigid and ineffective institutions.[53] Taleb's concern is that bureaucracy and other top-down efforts to protect against volatility and randomness are, effectively, the institutional equivalent of excessive hand washing. In his view, to avoid getting sick, we need to expose ourselves to germs. In other words, we need institutions that are less bureaucratic, less controlled, less orderly, and less risk-averse—institutions that embrace diversity, learn rapidly from failure, and evolve. Small governments love commerce, whereas large governments, he says, love war and produce deficits. Though an iconoclast, Taleb is one of an increasing number of complexity theorists who lament the oversimplification of social conditions that bureaucracy perpetuates. Contrary to early assumptions that scale demands bureaucracy, these writers set out to prove that bureaucratic hierarchies do not—and cannot—always work best in complex societies.

The Challenge of Deliberative Fundamentalism

Though it is awkward to admit this, the core assumptions even in our democratic polity and society are profoundly undemocratic: people are ignorant; attempts at participation will be plagued by group pathologies; large organizations require hierarchical, professional management to function efficiently. Such thinking comes not just from corporate economists or authoritarian sociologists. It is also painfully evident among many theorists of participatory democracy, whose vision falls short of imagining real institutions that promote thick forms of democratic experience and engagement.

Even deliberative democrats, who propound the value of citizen engagement through dialog, whether through public comment processes, referenda, or polls, limit citizen engagement to asking people how they *feel* about policies made by others. Deliberative democrats do not seek to leverage what the public *knows* to inform and shape how those policies get made in the first place. This thinking—let's call it *deliberative fundamentalism*—relegates citizen engagement to the realm of talk and participation to the realm of civil society. It does not accept that expertise is widely distributed in society and that people possess talents useful to governance.

In the deliberative democracy narrative, expectations of citizens are low. It is no surprise when so few opportunities actually exist for the public to participate directly in governance or that, when they do, government processes are opaque and inaccessible.[54] Despite occasional crowdsourcing of opinions in the public sector by e-petition websites like We the People or infrequent opportunities to craft or comment on draft legislation online in Finland or Buenos Aires, no government has systematically changed how it engages the public in decisionmaking.[55] Few governments invite citizens to contribute their expertise, as distinct from their opinions. The role of the public is to confer legitimacy on decisions, but not to inform or to make them. Deliberation is intentionally divorced from power and action.

Deliberative democratic theory encourages people to talk to one another about issues of public import in order to arrive at procedurally fair and legitimate oversight procedures. But there is little faith that citizens will affect outcomes.[56] They may feel, but they do not know.[57] This deliberative school of thought is heavily influenced by Habermas's discourse theory, in which the political, public sphere is seen as a conversation, but not one directly between the public and government. Rather, the dialog is mediated by television and radio. Civil society produces public opinion, which, in turn, is filtered through the mass media and influences the agenda of lawmakers. For

Habermas, civil society and institutional government are separated by a wide conversational chasm that only mass media can bridge.[58] The conversation and dialog among citizens is intended to interact indirectly with the policymaking process and with the practice of administration.

Among American political theorists who advocate deliberative democracy, James Fishkin is a leading proponent and interpreter of Habermasian political and social thought. Fishkin's signature experiment, *Deliberative Polling,* which he has replicated in a wide variety of geographies on issues ranging from European citizenship to Internet governance, measures how demographically representative groups of citizens change their minds after discussions with their neighbors and assesses how they receive and review balanced information about a topic.

The mechanics of deliberative polling are practical and powerful. In the archetypical model, a deliberative poll convenes a national random sample of the electorate in a single place. Fishkin then immerses the selected participants in the issues, by means of carefully balanced briefing materials, intensive discussions in small groups, and the opportunity to question the assembled experts and politicians. At the end of several days of working through the issues face to face, the experiment calls for polling the participants in detail and comparing the results against similar polls conducted prior to the start. The resulting survey offers, Fishkin says, "a representation of the considered judgments of the public—the views the entire country would come to if it had the same experience of behaving more like ideal citizens immersed in the issues for an extended period."[59]

Designed to prove empirically the value of informing voters, deliberative polling offers a model for using discourse to shape people's opinions. Fishkin, like other deliberative democrats, wants to refute the arguments of those who claim that voters are chronically incapable of learning. He demonstrates that people can be educated and will use their newfound knowledge to make decisions. The implication: deliberative practices can and should be instituted to improve the informational level of voters and, thus, the quality of their role as overseers of political decisionmaking.

However, such deliberative polling is about building a "respectful communicative process," not about achieving more effective outcomes per se.[60] Deliberative democrats are concerned with the legitimacy of the process, not its performance or outcomes. Many deliberative democrats, communitarians, and other strong democrats express greater confidence in the ability of citizens than those who focus on voter irrationality or participatory inefficiency. Many want to make deliberation a more regular part of political activities, but none see deliberation replacing day-to-day administrative decisionmaking practices.

Deliberation informs decisionmakers about people's preferences but is silent about citizen expertise. Deliberative polling focuses exclusively on opinions—on a more direct way of making civil society audible—but not on giving decionmakers greater access to public expertise. Its goal is to educate citizens, not tap their knowledge. Needless to say, deliberative democrats want decisions to be informed by hard facts, good science, and a general notion of truth. But they view the knowledge-creation process as something that happens off-stage among professionals.

Indeed, there is nothing inherent in Habermas's theory of deliberation, as interpreted by its leading theorists, to suggest that citizens have the competence to produce expertise or to be responsible for making decisions.[61] They simply do not possess knowledge-producing potential. The empirical literature on citizen deliberation focuses on measuring how many participated, who participated, who talked, with what tone, and how satisfied they were. It pays scant attention to their impact on policymaking or whether the resulting decisions were better tailored to achieving the desired outcomes. It counts as participation if people are asked to provide an opinion, regardless of the relevance of that decision for people's lives.[62]

Put simply, the deliberative sphere is a civic and secular congregation where people express opinions. It does not connect with action. Deliberation emphasizes the moral development of the individual, not the practical demands of problem solving or even agenda setting.

A more recent school of thought in political theory propounded by theorists such as Elizabeth Anderson, David Estlund, and Helene Landemore is "epistemic democracy."[63] Epistemics is concerned with questions of how knowledge is formed, and epistemic democracy focuses on the ancillary question of the practices by which people come to know and decide in a democracy. Epistemic democrats are searching for a philosophical justification for democracy as the normatively preferable mode of governance. They are trying to "save" democracy from epistocracy and rule by experts by reconciling the search for legitimacy and the demand for outcomes. If morality alone cannot justify democracy, they want to prove that democracy produces more efficacious outputs, not just fairer inputs. More than any other political philosophers, these thinkers take seriously the concept of citizen competence and the need for citizen participation.

Epistemic democrats go beyond deliberative democracy's emphasis on proceduralism, rejecting the notion that decisions in a democracy are good simply when they include demographically representative deliberative practices. They ask whether institutional arrangements of a particular type are better than others at obtaining the knowledge to solve particular types of problems. They explore how to design governing institutions to improve their ability to gather and make use of information. (By contrast, deliberative democrats focus more on procedure, on how decisions are arrived at using dialog and discussion, not on the resulting outcomes.)

But epistemic democrats tend to follow two rather traditional approaches to demonstrate the connection between democratic forms of participation and better decisionmaking: voting and large-scale deliberation.[64] In their view, democracy is superior not simply because it is procedurally fair, but because—when the judgments of large groups of individuals are aggregated through voting—democracy is mathematically more likely to produce the right result. In many ways, they start from the same assumptions about the relationship between citizen and state the deliberative democrats hold, but they arrive at a different conclusion related to outcomes.

The argument about voting being the better way to decide is rooted in Condorcet's Jury Theorem from 1785. When voters must decide between two options, they have a greater than 50 percent probability of being right—and if they vote independently and without undue influence, as the number of voters increases the likelihood of the majority being right approaches 100 percent. James Surowiecki popularized this understanding of what is often referred to as "collective intelligence" in *The Wisdom of Crowds* by pointing to experiments that show that when a large number of people guess the number of jelly beans in a jar or the weight of a sheep at a state fair, the average of those aggregated guesses is mathematically more likely to be correct. The law of large numbers works when applied to questions to which there is a right answer. By aggregating and tabulating large numbers of votes across a network, a group can be smarter than the sum of its parts.

The belief that the sheer number of people involved in making a decision improves the outcome has enjoyed longstanding popularity. The proposition that processes that involve many people are less likely to err than decisions made by a small number of people has its roots in the thinking of Aristotle, Machiavelli, and Spinoza, as well as the Marquis de Condorcet.[65] To democratic epistemologists, voting (as distinct from the coin toss, dictatorship, or oligarchy) is not simply morally right. It empirically improves the quality of decisionmaking because the aggregation of individual votes offers a solution to the "problem" of the average citizen's ignorance and irrationality. These same arguments about collective intelligence are used to justify the majority and supermajority voting techniques commonly found in a democracy. Contrary to Somin and those who posit that voting is irrational because a single vote does not matter, collective intelligence theory suggests that every vote matters because all votes together are more likely to be right.

This celebration of voting, however, again underestimates the competence of citizens. Unlike Fishkin, who attempts to prove the value of deliberation through empirical experimentation, Landemore and the other epistemic democrats are pure theorists, who deal in ideal types absent social contexts. Landemore writes about the concepts of voting (and large-scale deliberation followed by voting), but not about voters or voting working in practice to improve decisionmaking. Examples tend to be drawn from artificial settings. In practice, correlations between broadly inclusive, large-scale processes and outcomes may be negative or not even exist. She does not address the declining rates of trust or the perception of malfeasance and incompetence in modern-day democracies where voting is the norm. To make an argument powerfully, philosophers often have to shed complicating details. But by talking about citizens writ large as talented "blanks" without socially situated attributes, skills, or abilities, the epistemic viewpoint does not open up our thinking to the possibility of citizens as talented experts with substantive contributions to make to administration.

Henry Farrell, a political scientist, and his co-author and investigator, Cosma Shalizi, a statistician, also align themselves with the epistemic worldview. They are compelled by the argument that democracies know better. However, they are eager to improve upon earlier, more abstract assertions. Their sophisticated research seeks to model computationally how to improve upon current institutional arrangements, including hierarchies and markets, to tackle complex problems using public engagement. They, too, are grasping for evidence that more democratic ways of working lead to better outcomes and attempt to bring more mathematical and empirical rigor by running complex computer simulations to help prove why democracy works. Farrell is also undertaking parallel work designed to explore how to harness partisan sentiment to motivate rather than dampen productive engagement.

To date, however, most mainstream advocates of epistemic democracy have not yet tried to connect theory to any practical understanding of how people decide best, in what contexts, or about which policy or design recommendations. Their interest in collective intelligence is an important move toward embracing citizen competence, but their theoretical analysis still falls short of offering real practical wisdom for how to govern differently.[66] Even epistemic democrats, who claim to embrace the knowledge-producing potential of citizens, still view individuals as "intelligent" only by virtue of the law of large numbers. Endemic to the literature even of participatory democracy is the view that people are either congenitally incompetent or uneducated.

The notion that citizens are illiterate, pathological, inefficient, and, at best, capable only of talk divorced from action or of exercising an up-or-down vote on a topic runs throughout mainstream political theory. Institutional designs that take citizen competence seriously thus cannot grow easily out of mainstream political science, a field whose origins and raison d'être are deeply intertwined with professionalization and the sociology of institutions designed to elevate the status of a governing elite and exclude outsiders. Nor can a new theory of collaborative democracy that envisions citizens as partners with the state in action-oriented problem solving be cordoned off in a specialist discipline of politics or political theory; it is inevitably entangled with science and technology. These different fields of intellectual achievement, far from being distinct, are intertwined and becoming more so. Rather, it is our changing conceptions of expertise, engendered by new technology and developments in computer science, that are likely to have the most profound impact on the next stage of evolution of our political institutions.

The Technologies of Expertise

Where is the knowledge we have lost in information?
—T. S. Eliot, *The Rock*, 1934

Those who manage our economy and society need a more accurate and evidence-based picture of on-the-ground conditions. The widespread diffusion of sensors to collect large quantities of data—so-called big data—could help meet this need. Over the last two decades, we have learned how to create and store new kinds of content in vast quantities. In late 2012, it was estimated that there existed some 2.8 zettabytes (2.8 trillion gigabytes) of data; this was expected to grow to 40 ZB by 2020.[1] A 2011 McKinsey industry report said the volume of data generated worldwide is growing at a rate of approximately 50 percent per year—roughly a fortyfold increase since 2001.[2] Although many of these data-collecting devices, such as air and water quality sensors or speed trap and commercial cameras, are affixed in situ to the infrastructure of our communities, data collection mechanisms are increasing thanks to new sensors in our consumer electronics, including handheld and wearable devices. Our cars, fitness monitors, and mobile phones, just to name a few of the Internet-enabled devices that form part of the Internet of Things, often include a proximity sensor, accelerometer, barometer, light sensor, and more, making it possible for everyone to collect and stream—wittingly or unwittingly—massive quantities of data. We have also been creating more content than ever before by blogging and tweeting with abandon. We are collectively producing a Library of Congress–worth of information every day.[3]

There is a great deal of interest in unlocking vast treasure troves of big data to yield more granular insights about on-the-ground conditions. Much of the attention directed to "smarter" institutions has focused on this data that is passively gathered about environmental, biometric, epidemiological, and physical conditions.[4] Although enormous data inventories can help us to model whole systems dynamically, policymakers cannot use such raw information. The very technologies that are enabling us to produce more data and express ourselves creatively are, at the same time, diminishing the effectiveness of the institutions that manage the flow of information between government and governed. Counterintuitively, having more and more complex information about ever more complex problems impedes action.[5] We do not need more information. We need *human expertise* from individuals and from groups to curate and align data with decisionmaking.

Expertise—and expertise alone—can translate the available torrent of raw knowledge into "useful, relevant, informed, independent, authoritative and timely advice."[6] In parallel to the advent of new tools for visualizing and manipulating large quantities of information—the big data revolution—platforms, often called expert networks (or, informally, knowledge networks or talent platforms), that can fully automate the process of expressing, locating, and matching expertise within and across organizations have also started to become a reality. Although the tools and approaches are still evolving, it may soon become easy to identify with precision who knows what and match them to opportunities to serve. These technologies of expertise are unlocking the potential for crowdsourcing wisely and smartly, rather than crowdsourcing widely, and therefore hold the key for helping institutions find the right expertise to make sense of the ever-growing tide of data and make more effective decisions.

Why We Need Expertise

Human expertise is essential for translating information into actionable knowledge. Long before the era of big data, policymakers have sought evidence to support a position by cherry-picking information that was *reliable, specific,* and *relevant.* First, such curation helps to ensure that information is *reliable*. To many, the profusion of information via the Internet is anything but. Information can easily be distorted. New technologies for manipulating content allow users to edit their own digital text, pictures, and sounds and, increasingly, to collaborate in the creation of cultural bricolage.[7] Online information, says David Weinberger, is perceived by many to be a "victory lap for plagiarists, the end of culture, the beginning of a dark ages inhabited by glassy-eyed chronic masturbators who judge truth by the number of thumbs up, wisdom by the number of views, and knowledge by whatever is the most fun to believe."[8] Knowing what to rely on depends on access to people capable of making curatorial decisions.

Experts also help to ensure that information is *specific*. It is only through the lens of local knowledge—conditions on the ground such as topography, the layout of buildings, the progression of a disease—that information becomes something useful for decisionmaking. In the aftermath of a natural disaster, for example, it is vital to be able to get information from those who know the landscape. In the aftermath of a medical diagnosis, personal experience with the progression of the disease and its symptoms is vital to improvement. The UK's National Health Service continues to run a successful program known as the "Expert Patients Program" in which cancer sufferers and survivors provide support to one another to improve self-management of long-term health conditions with positive impacts on outcomes.[9] Experts are those with a unique perspective and the ability to curate through their own particular lens, helping us to see what is visible only through their eyes.[10]

Ground-level knowledge of cultural context and the sensitivities and priorities of relevant stakeholders, as well as of physical and natural conditions, is an important kind of expertise that emerges from practical experience and that bureaucrats often have a hard time replicating. "The man who wears the shoe knows best that it pinches and where it pinches, even if the expert shoemaker is the best judge of how the trouble is to be remedied," wrote John Dewey.[11] One of the key functions of lobbyists and interest groups is to provide this kind of specificity (albeit skewed in favor of a particular client). Specificity, however, often means different things to a decisionmaker in different contexts: obtaining missing facts and information that allow the right answer to a question to be ascertained; obtaining insightful options from diverse sources that allow the identification of alternative answers and interpretations; and getting at perspectives, such as those from people with situational experience, to fill identified gaps.

Decisionmaking also improves when information is *relevant.* Even if a statement is factually true and highly detailed, if it is not germane to the decision that needs to be made or if it is not available at the appropriate time and in a digestible form, then it is not actionable. Hence there is a need for meta-experience—the ability to sort and filter what is germane. With approximately 1.4 million academic journal articles published annually, there are many clever and innovative ideas swirling around, but in a format that decisionmakers have no time to read.[12]

There is a well-documented "knowing-doing" gap between the vast quantities of material penned by scholars who know, and the domain of policymakers who do.[13] Groups such as Footnote in Rhode Island and The Conversation in Australia, as well as inside-the-Beltway think tanks, "translate" the work of academics for the benefit of policymakers, but their reach is limited.[14] Although research consistently shows that bringing in good new information can help institutions make better decisions, all of us, including and perhaps especially government officials, are drowning in a tsunami of information that is not particularly helpful or actionable.[15]

This surfeit of content is making it progressively harder to align information with decisionmaking. That is why we still rely on *insiders* and *professionals*—along with a relatively closed, elite circle of interest groups and their professional research services—to curate, filter, and broker the data. A "profession" is shorthand for third-party training—training apart from the workplace, not "on the job" or in the field, but theoretical, rational, in-the-lab training focused on scientific principles of the discipline. Reliance on third-party credentialing by universities is the hallmark of professionalism today and is a highly prized emblem of middle-class social status and achievement. It is, however, merely a proxy for what decisionmakers need to know.

The value of this proxy—professional membership and university credentials—lies in the fact that it is supposed to reduce the transaction costs of searching for credible information about a person's expertise. As one commentator noted in the *Harvard Business Review,* "Americans have embraced degrees with a fervor generally reserved for bologna or hot dogs. Everyone should have them! Many and often!"[16] By common consent, a degree is excellent shorthand for ascertaining the level of an individual's expertise. All too often, however, it says little about that person's real-world skills and experience.

But governing requires the ability to curate quickly *credible, specific,* and *relevant* information, to make hard decisions. For instance, imagine if there were a way to identify those inside and outside of government who have led emergency responses, lived in the affected region, or who have experimented with open data or crowdsourcing to combat a scourge successfully. Imagine being able to connect quickly and get advice in the event of a pandemic. If only there were a directory of those with skills and experience in cyber-security; imagine being able to reach out to them in the event of an infrastructure attack. Imagine there were a way to push out questions about our greatest public challenges to networks of the most creative innovators and experts across institutions and organizations. Matching people to opportunities to contribute what they know could save time, financial resources, and perhaps even lives.

Identifying who knows what inside government—let alone outside—is hard. Within the bureaucracy, titles such as "director" and "manager" disclose little about lived experience. Because government projects are branded as institutional achievements, it is difficult to identify individual contributions. Yet decisionmakers are highly dependent on expert judgments.[17] Professionals inside an organization lack adequate access to diverse and innovative sources of expertise, especially in situations where situational awareness, local know-how, craft ability, disciplinary diversity, and the gritty "deep smarts" of lived experience are critical to good outcomes.[18] There are times when a whiz-kid hacker with no college degrees but lots of experience building software is more useful than a PhD programmer—or when a local resident in a particular neighborhood has more relevant know-how than a think tank scholar.[19]

Expertise, of course, takes many forms other than credentials and diplomas.[20] Some psychologists define an expert as someone who has devoted 10,000 hours or more to developing mastery.[21] Others discount the time invested in learning and, instead, point to native intelligence and genius—such as John Von Neumann's ability to instantaneously perform complex mathematical operations—as the hallmark of expertise. Harry Collins, a sociologist, and Charles Evans, a philosopher, offer a "periodic table" of forms expertise in *Rethinking Expertise*.[22] In their view, ubiquitous expertise is the kind of tacit knowledge we possess by virtue of the society and environment in which we live, such as the language we speak. This is the type of expertise language platforms like iTalki, which can connect a person interested in learning Spanish with a native-speaking partner, make use of most. By contrast, specialist expertise spans the gamut from popular knowledge (or what they call "beer mat" expertise, acquired by reading coasters from a bar stool) to the contributory expertise acquired within a broader knowledge community, such as the craft of an experienced scientist or a baseball player.

But expertise also includes the social skill to relate to others how an experiment was done—interactional expertise. An interaction expert is someone, like a science writer or a sports writer, who immerses herself with the specialists, much as an anthropologist studying a tribe of cavemen, without fully knowing how to replicate their art, becomes very good at describing it to others: telling instead of doing. This distinction is often most pronounced in the differences between academic and practical forms of expertise.

The effort to describe expertise more systematically and to take seriously the different forms of expertise that people possess echoes work done in educational theory by Howard Gardener, who famously propounded the theory of multiple intelligences—musical, bodily-kinesthetic, logical-mathematical, linguistic, spatial, intrapersonal, and interpersonal intelligence. All of us have all of these intelligences, but no two of us have the same combination in the same measure. Some of us are doctors; others are musicians and poets. Some are talented at painting a picture; others at painting a fence. Some are introverts; others, people persons. Some have the ability to play basketball well; some can do a play-by-play recounting of the game but cannot dribble. This variegated know-how exists among those with or without university credentials, inside or outside of government, with a theoretical or a practical bent. Each one of these dyads represents a source of previously invisible expertise that technology is making more searchable.

Four Ways Technology Is Changing Expertise

People searched for experts long before the advent of computers, of course. But with advances in information retrieval technology, including natural language processing and the large-scale availability of digital traces of knowledge-related activities, the processes of matching the supply of expertise to the demand for it within organizations is becoming more automated in the twenty-first century.[23] Expert networking platforms—also called people search, expert discovery, expertise retrieval, expert finding, expert profiling, and e-expertise tools—use software and associated algorithms to make expertise manifest. HeyPress is a website to search the web for journalists matched to one's needs. Gradberry does the same for software programmers. CyberCompEx is an expert network for cybersecurity practitioners. These matching tools are proliferating in every field.[24] A variety of still nascent techniques that measure expertise in different ways are making it easier to organize and catalog who people are. Experiments using test-bed communities and training datasets are investigating how to learn who knows what within an organization.[25] There are also widely available, open, and diverse sources of data that are searchable *across* institutions. We are on the cusp of an expertise revolution, not just an information revolution.

Generally speaking, these tools incorporate profiles showcasing what people know. Using a combination of data scraped, for example, from social and professional networking sites such as Twitter and LinkedIn, from public sources such as website profiles, grants, and publication records, and from information provided by users themselves, these tools organize data into easily searchable directories. In addition to data sources affirmatively contributed by people to websites, some tools also mine the "digital exhaust," or unwitting traces of activity such as browsing habits from websites or keywords from documents. The directories then sort people on the basis of criteria such as reputation, credentials, skills, and experience.

These automated expertise retrieval tools solve a commonly confronted problem—the need to *segment an audience*. Academics, for example, need to know who is working in a field in order to construct a productive and original research agenda that does not overlap with that of others and that tackles important gaps in a field of study. In fact, they are among the primary users of expert networking tools. VIVO, an academic expert discovery platform, lets its 100,000 members in 100 universities search for grants, peer reviewers, and potential collaborators; its database can help scientists and authors avoid conducting duplicative research.[26]

Because expertise discovery often works by inferring expertise from the association between people and documents—such as by counting papers authored or cited, grants submitted or awarded—academia, where the currency of the realm is the journal article and the grant proposal, has been a forerunner in the use of expert networking technology. Since corporate systems tend to be proprietary whereas academic ones are open, it has been easier to develop expertise retrieval technologies in the academic space. Purdue's Indiana Database of University Research Expertise searches experts across the state's four major universities (Ball State, Indiana, Purdue, Notre Dame), draws on faculty homepages, descriptions of National Science

Foundation–funded projects, faculty publications, and supervised dissertations—all open sources—as well as dedicated profiles. Profiles include a predefined taxonomy of expertise as well as the ability to add new keywords.[27]

The Los Alamos National Laboratory developed EgoSystem to go one step beyond academic research discovery; it keeps track of and supports outreach to former employees in the private sector for purposes of building public-private collaboration and partnership. *EgoSystem* combines sources of data from biosketches, which include details about the places and topics a person has studied, with that person's social media output and information about him or her from other academic networking websites such as *Mendeley,* which millions of people use to keep track of the PDFs of journal articles they read.[28]

Tracking the outcome and impact of research is another use for such tools. The State of Sao Paolo in Brazil, which accounts for 35 percent of the country's GDP, dedicates 1 percent of its budget to research funding under the state constitution. As part of the funding submission process, applicants are legally required to complete a Google Scholar profile to facilitate the tracking of investment impact as well as catalyzing greater collaboration among the country's academics.[29] There are myriad additional needs for tools to match people on the basis of desired expertise. There's gold in knowing who knows what. Health Tap matches doctors to patient questions. The company advertises a catalog of more than 64,000 medical professionals. Would-be patients can broadcast a medical question and either invite a reply from a doctor through an open call or search the answers to earlier questions. The company hopes that searchers will become paying subscribers who then may use the computer system to direct questions to specific doctors. Participating doctors have profiles that show their credentials, publications, and specialties, of course. Those profiles also show the previous questions they have answered and how people rated their responses. They list the doctor's rating by other doctors and by clients on the system. The site can also be browsed by topic, such as pediatric rheumatology or allergies. The company boasts thousands of lives saved and millions of questions answered.

It has long been understood in business, especially electronic commerce, that segmenting and targeting individual audiences is essential because segmentation improves sales performance over aggregated models of marketing.[30] Segmentation in the business context enables price discrimination.[31] Real-time information about consumer preferences, behaviors, and intentions can inform messaging content and delivery.[32] Studies have shown that targeted advertising produced 2.7 times more revenue than nontargeted;[33] it also boosted employment, raised innovation, and increased economic productivity.[34]

Statistical techniques were previously enlisted to do the work of inferring age, marital status, demographic, income, and location information and roughly predict behavior in the aggregate. One early experiment called "White Rabbit" developed profiles of people based on their conversations.[35] Building on these earlier statistical techniques, the advent of "big data," has made it easier for companies to mine vast quantities of transactional and behavioral data from myriad sources about a single individual to develop better predictive models, enabling ever more granular micro-targeting.

When someone searches online for cars or golf clubs or reads online reviews about them, sophisticated websites track that behavior, often using a cookie on the customer's computer; they infer from the context that the searcher is interested in buying those products and serve up the appropriate ads recommending products that are adjudged to be most relevant.[36] The scope and depth of targeting depends on the industry, consumer privacy expectations, data availability, and the feasibility of micro-targeting.[37] Sometimes these techniques work too well, as when the predictive analytics unit of Target infamously (and correctly) guessed that a teenage woman was pregnant and targeted her with relevant product advertising before she had told her parents the news.[38]

Increasingly, companies like Facebook and Google, which have accrued massive quantities of personal information about their users and how they behave online—on the web, on mobile devices, and in the physical world (Facebook check-ins, Google map usage, etc.)—are learning to micro-target ads with even more precision to exactly the desired demographic, using proprietary modeling and segmentation techniques. There is no shortage of PhDs working on the algorithms for improving online ad targeting to predict more accurately what a customer will buy and in which contexts and circumstances and how best to exploit what is known about people to match them to products.[39] Based on a viewer's browsing history or behavior on multiple platforms—their so-called data (or digital) exhaust—these companies, especially when they combine their data with that of data aggregators like Acxiom, can guarantee that an advertiser is reaching the desired population and increase the likelihood of converting a viewer into a buyer.

Although profiling has become part of everyday commerce, these applications of user segmentation and targeting make some people uncomfortable. Companies' use of highly granular data about an individual's behavior and attributes triggers questions of personal privacy.[40] Social media sites, browsers, and other common tools are set to guess at gender, age, and other attributes that become part of the default settings of the program, which are then used to determine which ads will appear in the sidebar. Google ads, for example, run off automatically determined guesses as to gender, age, and interests, which the software configures based on browsing habits. Although Google has made these easy to view and change, web browser settings are remarkably hard to find.

But if we can develop the algorithms and platforms to target consumers, can we not also target citizens for the far worthier purpose of undertaking public service? LinkedIn is partnering with Volunteer-Match to match people to opportunities to volunteer.[41] Why not match citizens to opportunities to participate in governing? Why not investigate strategies for helping government and citizens to understand each other better? How can we not at least explore facilitating the opportunity for people to participate in democracy and civil society in ways that speak to their talents, interests and abilities? Of course, we must do so in a manner consistent with the values of privacy. But just as an advertiser wants to be able to target the right audience with ads matched to their interests, a policymaker should be able to invite those with something to contribute. Such targeted crowdsourcing would not prevent anyone from signing up to participate on their own, but would make everyone aware of the opportunity and unlock a whole new dimension of participatory democracy.

In centuries past, professional membership provided an alternative to aristocratic heritage and patronage—a middle class escape into the world of status, power, and wealth; today the new technologies of expertise portend a similar social transformation of traditional professions and credentials. The technologies of expertise can be employed to reorganize and restructure administrative institutions by making the expertise of all people—not only an elite with educational credentials—more visible, searchable, and useful for tackling public problems. The Internet is democratizing expertise and opening up the possibility of connecting anyone to an opportunity to engage in the life of our democracy that speaks to his or her talents. Traditionally, as we have seen, expertise was associated with educational credentials and professional affiliations. More credentials are, undoubtedly, a hallmark of greater education. However, credentials convey only a tiny slice of the skills and expertise useful for public problem solving.

The Internet is changing our relationship to expertise in four key ways. First, technology is making a broad range of expertise—including more forms of craft knowledge as well as skills, experience, interests, and credentials—more visible, independent of institutions. Second, it is permitting expertise to be identified by means of manual and automatic data collection, thus enabling an infinite diversity of expressions of expertise based on different traits and characteristics. Third, it is proliferating the number and type of expertise markers, including the shift from certification to badging. Fourth, Internet technology is making it possible to quantify expertise differently and more diversely than before. New technology is helping to strip "expertise" of its elite connotation and render it a more neutral—and actionable—description of talent and ability.

Making Expertise Visible

Expert networking tools are making it possible to *display* and *make searchable* the talents of diverse populations. Anyone can now call attention to their experience, skills, and interests with myriad new markers of expertise. If I take a course in programming on Udacity, Coursera, or Lynda, my certification will be visible online for others to see. But unlike a traditional university transcript, this certification is easily displayed and searched, not controlled by any gatekeeping institution. From posting a résumé on LinkedIn to earning a badge for honest salesmanship on eBay and Amazon, new technologies of expertise (also called talent platforms) are making it easier for me to showcase what I know—and for others to target me on that basis.

Expertise markers go well beyond credentials. Many sites boast leaderboards to showcase people's accomplishments.[42] Platforms that display a user's contributions on the screen positively correlate with increased participation. Seeing one's achievements reflected back through an interface—on a leaderboard or through badges or a point system—reinforces one's inclination to participate.[43] Joan Morris DiMicco (formerly of the MIT Media Lab and now at IBM Research) also studied what happens when groups see their own behavior reflected back.[44] This is what she called social translucence or social mirroring: greater mutual awareness of actions that can result in more effective coordination as a group. It is not a big leap to imagine the impact on both individual and social consciousness of seeing indications of one's own skills and abilities (and those of other people) reflected in new visual markers on the screen.

The Internet is also transforming how we *define* expertise, expanding our definition of the term to capture more varied forms of craft knowledge, including *skills, experience, and interests.* For example, although LinkedIn users complete a profile of their education and work history, including traditional credentials, the social network also invites users to describe their skills, automates the ascription of skills to a user, and asks the members of a person's network to confirm and endorse these ascriptions. A recommendation is a comment written by a LinkedIn member to endorse a colleague.[45] LinkedIn recommendations let third parties vouch for your skills without requiring any action (or responsibility) on the part of LinkedIn.[46]

Many sites let users present their experiences rather than credentials. On the Khan Academy learning platform, an online educational institution, students receive points for every video watched and exercise completed, as well as extra points for demonstrating mastery of concepts. On Coursera, an online learning platform that delivers university-style lectures by professors at leading universities, students can show how many courses they have completed in Python or C+. Although taking a course on Coursera does not attest to their ability to do anything, posting a video of themselves on wikiHow, YouTube, or Lynda teaching someone else how to tackle a programming challenge is evidence that they possess knowledge and a skill—evidence further reinforced by page views and downloads by others.

Markers of expertise are myriad and awarded from a variety of platforms—many of which are online only—that do more than offer courses. Expertise may be demonstrated in the form of diplomas earned, skills or interests followed, experience (including work) completed, milestones accomplished, tasks and challenges undertaken, and time spent on an activity. Now users can read ten books about programming but also have a designation from Amazon showing that they did (along with metadata about how long it took). When they apply this new knowledge by writing a software program, they can demonstrate new mastery by saving the software they have written to a public repository like GitHub (known as checking in code) or answering questions from other programmers on a question-and-answer site such as Quora or Stack Exchange.[47] When they complete or even win challenges on DataDriven or Kaggle, expert networks for data scientists, they can earn the ascription of "kaggler" or "master."

In the newly emerging collaborative, peer economy, experience is fast replacing credentials as the way to measure mastery and expertise. People are sharing goods and services without the use of large, centralized intermediaries. This practice gives rise to the need for ways to certify trust. EBay started the entrepreneurial trend by making it affordable for anyone to create a market for the contents of their basement. An Ebay member with a Yellow Star rating has participated in ten to forty-nine auctions, while a Silver Shooting Star, signifies a million or more transaction ratings. Now Airbnb enables homeowners to rent their homes and rooms to visitors; and tourists can trust (or reject) an Airbnb rental because of the ratings of other users. Uber provides the software to enable anyone in one of more than a hundred cities to become a livery driver (local regulations permitting) and riders to hail a car at their convenience. Side Tour glassblowers, dumpling makers, and craft beer brewers offer unique experiences to tourists. Etsy is a marketplace for individual handicraft makers who want to reach a larger market. Peer Academy in Australia provides the space and training for anyone—through peer-led learning—to train others. Without any judgment as to the viability of plying one's trade in the peer economy or the ultimate impact on workers and consumers, these entrepreneurs depend on software platforms to manage both their recommendation and their relationship metrics (whether they had happy buyers) and their *experience* (how many times they rented a room or conducted a tour).

In some contexts, interests, rather than skills, experience, or credentials, may be a better proxy for expertise. If a user loves sports, she is more likely to know the ins and outs of the game. More precisely, although a baseball player can throw a curve ball, an ardent fan—the one with interactional expertise—might do a better job of calling the plays. In fact, the fan (and surely the blogger) might know more about statistics than the player. Similarly, a patient with a chronic illness may have more expertise than a highly credentialed doctor about certain aspects of that disease's care and treatment because he is more invested in finding something that works. Certain kinds of expertise require actual doing, but astute observers motivated by passion are often better positioned to communicate what they know.

Business professor Panos G. Ipeirotis has been running experiments that test whether interest is a useful proxy for expertise. While on sabbatical at Google, Ipeirotis set out to develop approaches to "crowdsourcing all the knowledge in the world." To that end, he designed Quizz.us to explore whether it was possible to crowdsource in a predictable manner with knowledgeable users. Using Google keywords, Ipeirotis advertised to potential participants the opportunity to participate in a fact-based knowledge quiz. For example, if you were on WebMD, you might receive an ad asking you to take a health quiz about disease symptoms. He wanted to see if he could get good users, not just clicks.[48]

By targeting a request to participate in health quizzes to Google users browsing on WebMD or the Mayo clinic, the Quizz.us project was able to increase the rate of participation by a factor of ten. In addition, volunteers from the targeted pool were also 20 percent more likely to answer correctly than those who found the quiz elsewhere. The self-selected pool also answered more quickly than those who were paid on Mechanical Turk or oDesk (two websites for farming out paid task-based work) to answer the same questions. Interest seems to be a powerful proxy for expertise. The WebMD surfer is better at answering health questions than are paid participants.[49]

Automatic and Manual Sources of Data

The new technologies of expertise also make it possible to *cull and collate information, manually or automatically,* to create a profile of what someone knows. Credentials, experiences, skills, and interests can be entered by hand when people respond to surveys or complete a bio-sketch. Npower is a nonprofit organization that provides IT services and support to other nonprofit organizations with an emphasis on social good. Npower's "Community Corps" program connects skilled individuals looking to volunteer their time with organizations that need specific skillsets. Volunteers fill out a form, report their skills and abilities during the registration process, and are screened by an Npower employee before being assigned to a task. Self-reporting is the linchpin to the process.

Form-filling is not new. It is as old as paper and bureaucracy. Manual data entry done online, however, is of interest for four reasons. First, "smart" forms, which know which question to ask based on previous responses, minimize the task of responding to questions, shorten the form, and create an incentive for completion. Second, the implementation of powerful new taxonomies—such as drop-down menus to describe different domains of knowledge or relevant skills—for representing expertise results in structured or semi-structured data that can easily be manipulated and combined with other data. The forms reflect the design of the underlying schema, which map knowledge areas and capture attributes of expertise. Third, online form-filling results in data about expertise on a large scale. By offering the prospect of professional advancement, web platforms like LinkedIn give users a strong incentive to complete the forms correctly and thoroughly in response to specific questions and to keep their profiles up to date. These structured and heavily interlinked datasets open up new opportunities for advances in expertise finding.[50] Finally, the manual data entry process generates an artifact—the profile. The profile is a document that represents the person's expertise and can be more easily searched than expertise from multiple sources can be mined on the open web.

Nonetheless, it is not always straightforward to ask people to answer questions about their skills, experiences, and interests. It demands an investment of time and effort that people are often reluctant to make and, once made, reticent to keep up. University administrators and deans complain about the teeth pulling required to get faculty to do annual updates about their research. Often the challenge is knowing how to phrase the questions to elicit a response. At a seminar at Oxford University, I asked participants to interview one of their fellow classmates at the table and, using pencil and paper, list a skill they had learned that the colleague possessed. Among the items listed were "light-touch" academic writing, knowledge of a disciplinary literature, mixology prowess, and hummus making. No two people understood the question the same way.

Trustworthiness in self-reporting is also a challenge. The light-touch academic writer might have exaggerated her prose style and the mixologist overrated his martini. Self-reporting can be inaccurate, by virtue of under- or over-selling. Online forms, at least, afford the opportunity to explain the questions, show other model answers, and normalize responses to account for fluctuations.

In some cases self-reporting—the mere act of presenting one's skills, experiences, or interests for the first time—can be a powerful incentive for participation. The platform Patients Like Me lets patients post information about their state of health, the drugs they use, and side effects they experience; a study found that—in a highly active community that is committed to the process of sharing information—some deceased users' spouses entered information after their deaths.[51] People want to be asked and want to share if given an interface through which to do so.

Expertise can now also be automatically scraped from third-party sources of data. Type the name of a professor into Google and among the first search results is a list of publications provided by Google Scholar, a specialty search engine within Google. According to Anurag Acharya, Scholar's creator, who developed the project as a volunteer "20 percent" time initiative—Google's program to allow its employees to spend one day a week on side projects—Google Scholar is designed to identify who has written what.[52] Google updates authors' publication lists by regularly scraping and combining different publication databases. Some private companies mine corporate email to determine who is talking to whom and about what as a means to "back into" the creation of catalogs of expertise. In 2006, for example, IBM launched SmallBlue, a system for mining both the content and the metadata of email to reveal the expertise that exists within an organization.[53]

There are companies that specialize in extracting expertise from a wide array of openly available databases. Usually, the resulting services are closed, proprietary (and lucrative) directories of expertise. Web of Science, a Thomson Reuters company, mines publication records, including books and journals, as well as patent data. It also looks at patterns in co-authorship to compile and visualize an academic's network of co-authors. Then it sells access to its research database, which scholars can use to find publications, but also to do visualizations and analytics that show who is connected to whom for purposes of finding research collaborators, identifying trends, and uncovering related publications. Like LinkedIn, these platforms take advantage of distributed data processing techniques and the ability to use multiple servers working in parallel to crunch large quantities of information from heterogeneous sources in order to promote research collaboration.[54] Symplectic is one example of an institutional knowledge management tool for universities. It populates public research profiles of hundreds of thousands of academics to automatically produce biographical information about an institution's researchers. Such research networking tools are especially popular in the biomedical sciences as a result of the rich availability of open publication data.

These automated platforms are also being used outside the academic environment. Relationship Science, a New York–based business intelligence company, combines machine and human processing to create a database of three million influential people and their organizations, including their work history, board connections, deal history, education, nonprofit donations and affiliations, political donations, personal interests, creative works and awards, business relationships, and relevant familial connections.[55] These platforms increasingly draw on new sources of data. A Klout score, for example, is a vanity indicator that translates all of a person's activity on social media into a single number. It is intended to measure influence in the world based on how many people retweet and quote what a person says. Klout boasts over five hundred million scored profiles.[56] If a policymaker, for example, wanted to encourage compliance with a new law, he might use Klout to identify and ask the most influential people to promote the initiative, in turn, to their own networks. Klout scores are reminiscent of the MacLandress Coefficient, a fictional metric conjured up by John Kenneth Galbraith in his 1963 novel as "the arithmetic mean or average of the intervals of time during which a subject's thoughts remain centered on some substantive phenomenon other than his own personality."[57] Although Elizabeth Taylor rated three minutes, Richard Nixon scored only three seconds. Politicians tend to have very low scores in Galbraith's world. What was then a "bug" and the butt of satirical humor in Galbraith's novel is a "feature" on Klout. The score measures a person's influence, which in turn could accelerate the process of identifying and targeting useful social amplifiers.

Especially with the discovery now by data scientists at Imperial College of how to pinpoint who influences whom on specific topics and in specific geographies via social media—the *direction* in which data flows—expertise tools can be used in a more targeted way to identify influencers within specific populations.[58] For example, the Obama administration enlisted the help of more than a hundred celebrities from Hollywood, music, and professional sports in 2014 to push the message to young people that they needed to sign up for health insurance under the Affordable Care Act.

There is also a hybrid between automating the making of an expertise profile and manually inputting information. VIVO's tools for researchers automatically populate a profile with information that the user corrects as necessary. Credentials culled from the web are often historical and must be annually edited for timeliness and relevance.[59] On ResearchGate, a network of five million scientists intended to help find collaborators and co-authors, members are encouraged to supplement their automated profiles with raw data and even failed experimental results in order to prevent the repetition of scientific research mistakes.

Shifting the default rule—that is, asking users to correct rather than input their descriptors—is attractive to people who would rather undo wrong information than take the time to fill in blanks.[60] LinkedIn and other platforms use "recommender" tools to prompt third parties to identify and rate a person's expertise. Expertise may be indicated, not by what a user inputs or from data about that person, but by aggregating ratings from those with whom that person has done business. With the ability to combine and recombine data inputted manually and scraped automatically, data science advances are making it possible to create new expert networks rapidly.

From Certification to Badging

The web is not only enabling the identification of new kinds of expertise from various sources; it is also giving rise to *new certifications* of know-how by actors other than traditional universities. Certification often appears in the form of a digital badge: imagine the graphical representation of a merit badge on a web page with embedded metadata explaining the certification. These are easily searchable by someone trying to target unique forms of expertise. A digital badge is an online record of someone's achievement by the community that issued the badge and of the work required to earn the badge.

Because the badges are digital, unlike a diploma, they are information-rich and verifiable sources of evidence about a person's achievements. More than a decade ago, Creative Commons, a nonprofit organization that enables the sharing and use of creativity and knowledge through free legal tools, popularized the idea of metadata-rich badges by creating a series of icons to represent different copyright licenses. The picture is simply a graphical representation of a much longer legal document. The widespread use of such licenses, coupled with the popularity of online video games where people create and play through graphical representations of themselves known as avatars, has paved the way for creating graphical representations of credentials akin to digital merit badges.[61]

In the offline world, one's experience is almost always self-reported (often ad nauseam at cocktail parties) and backed up with reference to hard-to-search diplomas and transcripts doled out by accredited universities and other institutions. Diplomas say very little about the work undertaken to obtain them, and transcripts, although proxies for performance in a course, offer little insight about a person's skill level. But in the online world, CodeAcademy *certifies* that X has taken a course in Python. Audible *proves* that X listened to the first half of a book. Amazon Kindle *validates* how fast X read the other half. The Internet puts Boy Scout achievement badges on steroids by enabling almost any online or offline community to *award and certify* a credential indelibly. Instead of only teaching graduate students at their home institution, faculty can offer their courses online to professional learners who, if they complete the same requirements, receive a permanent digital badge. This is more than simply the digital equivalent of the laser-printed diploma. The badge is numbered and escrowed with a third-party service provider that maintains and authenticates the integrity of the badge and makes it searchable and findable by others.[62]

What makes a Harvard diploma or a Cisco router maintenance certification valuable is the blue-chip quality and longevity of the certifying organization. We know that Harvard will be around forever, which makes its credentials more valuable than a certificate from "Frank and Morty's School of Law and Cosmetology on the Lower Level of the Seven Hills Shopping Mall" (made famous in a canonical article in the *Yale Law Journal*).[63] Within any given profession, one's life chances are strongly influenced by the prestige of the institution from which one receives credentials.[64] But with digital encoding, a badge or certificate awarded online by a website, although it does not carry the value of a Harvard diploma, can have at least the same permanence at a significantly lower cost and with greater discoverability.

These "nanodegrees" may actually say more about a person's skills than a transcript, especially to a busy hiring manager or policymaker looking for people with real know-how. Depending on the person doing the searching, a badge attesting to someone's practical programming or auto-repair skills may be of greater use than a traditional diploma. Even universities that award traditional paper certificates, such as New York University, are also issuing digital badges. Their graduates are demanding a more portable, visible, and permanent certification of accomplishment that can be affixed to a CV, shown to an employer, easily read by a robot (which are sometimes the first ones to screen job applications in large organizations), or browsed online.

By making it possible to take a single class or single module online (and pay for certification), online learning providers offer students a trusted and verifiable accreditation—a digital badge—without the overhead of enrollment in a full degree program. As automation puts pressure on workers to innovate, there is growing demand for ongoing, affordable learning and human capital development. More and more companies, from McGraw-Hill and Pearson to new entrants such as Amplify and Cengage, are either offering learning or the tools to create alternative learning environments. Universities are slowly beginning to follow suit by "unbundling" their course offerings in order to stay competitive.

DuoLingo, the language-learning software developed by Carnegie Mellon computer scientist Luis von Ahn, is now challenging the monopoly held by the Test of English as a Foreign Language (TOEFL) by offering an app for certification of English-language proficiency for one-tenth the cost.[65] Using DuoLingo enables the applicant to provide certification to a prospective school or employer rather than relying on TOEFL to distribute the test score. It is a short hop to imagine test takers checking a box to make themselves searchable by American graduate schools automatically.

Aiming to increase the "shareability" of badges across learning communities, the nonprofit corporation Mozilla, the makers of the Firefox web browser, launched a program called Open Badges to make it possible for any group to issue a badge and for users to combine multiple badges from different issuers and share them with employers or on the web. As Mozilla writes about the program, "Whether they're issued by one organization or many, badges can build upon each other and be stacked to tell the full story of your skills and achievements."[66] The Lego-like infrastructure for democratized credentialing is in place, but it is still waiting for more people to recognize and use it. The MacArthur Foundation is investing in this badging infrastructure, betting that an open system for digital badges will improve educational outcomes in U.S. cities.[67] With the cooperation of several mayors, MacArthur is sponsoring the creation of platforms to enable any education provider, from after-school clubs to nonprofit organizations, to offer a badge that is searchable across organizations and cities to give learners, especially high school students, a way to distribute evidence of their accomplishments.

Badging takes place in work environments too. TaskRabbit, the marketplace for twenty thousand task workers whose services can be contracted to do anything from running errands to unpacking boxes to designing websites, awards badges to its "elite" TaskRabbits. Not unlike the Power Seller on eBay, badges denote the highest rating for service and performance as evaluated by *customers,* not by the platform provider. Kaggle is the world's largest online community of data scientists, who compete with each other to solve complex data science problems. Kaggle creates profiles for scientists whose skills are rated as being in the "top 0.5%," based on the answers they submit to Kaggle client competitions. Kaggle uses these profiles to connect scientists directly with Kaggle clients who are seeking specific skills.[68]

With electronic certification and badging, individuals no longer have to rely on a handful of universities, which are, in turn, ranked by an even smaller cadre of publications like *U.S. News and World Report.* Control over certification is distributed to an unlimited number of other platforms such as TaskRabbit and Coursera. From the perspective of those seeking to crowdsource smartly rather than widely, however, badges are making expertise more searchable—which potentially enables leaders and managers to find and match people to opportunities that will make use of their unique skills and accomplishments.

Quantifying Expertise Differently

In addition to democratizing the certification process, the Internet makes it possible to *quantify expertise differently* and diversely, to search for those with particular kinds of know-how. New algorithmic approaches to mashing up data from different sources, such as publications, tweets, and digital badges, enable the creation of an almost infinite number of expertise directories.

Citation networks like Google Scholar or CiteSeer, designed for academic communities, track trends in scholarship by measuring citations.[69] When we can "see" that people in biology are citing people in physics, or that formerly unrelated researchers in computer science are starting to cite those in public health, we can literally see the emergence of new fields, like biophysics and medical informatics. A citation network like Google Scholar is essential for helping people—in this case journal authors—establish their expertise in scholarly environments. They adopt a specific algorithm that weights heavily for publication statistics—not social media mentions, for example.

Technology is also changing *how* different kinds of expertise get quantified as part of a larger trend toward *quantifying and measuring ourselves*. To date, this trend has emphasized personal measurement for improving health and well-being—measuring heart rates and other medical indicators, for example. But similar technologies could also enable more precise measuring of what people know. Lumosity, a popular commercial platform for improving neuroplasticity using memory games, brain teasers and puzzles, is an early example of such data-driven approaches to calculating cognitive capacity. In every case, new technology is accelerating measurements rooted in real-world activities rather than laboratories.[70] These algorithmic techniques allow new measurements of who knows what by *quantifying links to others, credentials,* and *experience or demonstrated success.*

Measuring links between people using relational data is one way to quantify expertise. We can search for who knows what using the names of people we know, people who know us, and people who recommend us. For example, the number of people in a social network I am connected to can be mapped and graphed as a way to demonstrate my influence and standing.[71] On Facebook, expertise is often implied on the basis of whom one knows. "Friendship" with specific individuals or groups of people with reputations in a specific community creates a presumption of expertise by association. These social associations are one way to pinpoint expertise.

In one study, participants posted a question to their Facebook social network and, in parallel, searched the web for the answer. Over half the participants received responses from their social network before completing their web search.[72] Similarly, on Twitter, to follow someone is an implied declaration of interest in that person, even if the act of following comes from a desire to keep one's enemies close. Being followed by people implies a recommendation rooted in the status and skills of the recommender. In such a relationship-based recommendation system, my status would be measured by who knows me, such as on LinkedIn, where reputation results from being connected to people of relevant and reputable standing and accomplishment. Because Twitter is frequently topic-based, searches between people who follow one another often also turn up expertise, not simply association. In such systems, the assumption is that the willingness to declare a personal relationship publicly results from a history of positive performance and translates into a useful proxy for relevant expertise. Because of the relative technical ease of measuring tweets, retweets, Twitterrank, likes, follows, comments, rating and ranking scores, and other indicators of approval on social media, these platforms offer an ideal and flexible testing ground for new techniques for expertise retrieval.[73]

Relationships can also be measured through citations, such as on Google Scholar. Citations are simply a special category of the tie between nodes on a network. This need not be limited to academic citations. Social media sites sometimes measure influence based on retweets or links back to a person's online postings. These kinds of measures of a person's ties, as calculated by mentions in social media or quotes in articles, drive sentiment analysis, the techniques that network scientists are pioneering to detect patterns in who is talking about whom. Sentiment analysis, social filtering, and other such strategies are being applied to data about who knows what, such as citations or tweets or words in biographical profiles, to make it possible to search for people more effectively.

Expertise can also be measured by assessing how many people have viewed, read, and downloaded one's academic papers, not only cited them, which is a much smaller number. The academic networking site academia.edu uses page views to create its bibliometric. On a competing platform, the Social Science Research Network, scholars upload their papers, and expertise then gets quantified by the number of times they are downloaded. Academic hiring committees regularly use citation, page view, and download scores to gauge the expertise and standing of job candidates.

All of these techniques are variations on a single theme in data science: measuring and quantifying expertise based on the strength of ties in a network. Recommendations can come from those we know well or those with whom we have interacted and done business. For example, we rate books on Amazon and movies on Netflix with stars or points. This kind of third-party rating of content can also apply to people.[74] Net promoter systems, which ask for approval or disapproval of a doctor or another service provider, are often the most powerful measure of expertise and effectiveness.[75]

Some recommendation systems are rooted in a personal association where the recommendation derives from an intimate knowledge of a person's skills, ability, and experience, such as when colleagues recommend each other using the LinkedIn recommendations feature. Similarly, on the now-defunct site Vow.ch, participants gave assurances for one another rooted in strong ties. Strong ties are considered instrumental for influencing online and real-world behavior in human social networks.[76]

Recommendation systems are not original to the Internet. Jobs and universities have always benefited from recommendations as a way to prove expertise. Any form of "old boy" network relies on introductions and mutual endorsements. Such recommender systems are not simply nepotistic. Rather, those who know about something are simply more likely to know others who know about the same thing. In other words, the expert heart surgeon is more likely to know other excellent heart surgeons, just as the jazz aficionado will know others with similar taste in music.

It was this contention—that people who have special expertise or attributes will tend to know of people who have more of that expertise than themselves—that Eric von Hippel and his colleagues set out to test empirically in their work on pyramiding (sometimes called snowballing).[77] It turns out that pyramiding—asking a known expert to identify people who know more than she does about a topic (and repeating the process with each successive recommendation)—is an especially good way to find out who has rare forms of expertise within large but poorly mapped knowledge domains.[78] Modern relational data science transforms and accelerates the process of recommending into an objective and quantifiable metric. It also reduces the role that gender, race, and formal organizational authority play in determinations of expertise.[79]

Furthermore, recommendations can be made even about people encountered only briefly in a single transaction—weak ties.[80] On the e-hailing taxi site Uber, passengers and drivers rate each other based on their mutual courtesy during a single ride. My rating of the driver, combined with the ratings of other passengers, yields an overall rating for that person as a taxi driver. If that driver receives a consistently bad score, Uber may stop doing business with him. On eBay, buyers and sellers rate each other for performance. Future buyers rely heavily on reputation when deciding whether to transact across long distances with an unknown seller. Reputation points are gold on eBay, and without them the ability to sell is severely diminished.[81] Tutoring sites such as WyzAnt and Ohours, which offer a platform for organizing "office hours" for people to teach each other, rely on personal ratings of someone's level of knowledge and skill at imparting it. "We are all being rated these days," comments Delia Ephron. "Doctors, professors, cleaners, restaurants, the purse repair store I looked up on Yelp this morning. . . . Who wants to hear what anyone thinks of you? That's why talking behind your back was invented. . . . I have long been done with school. Nevertheless, it turns out, I am going to be getting report cards for the rest of my life. Remember your permanent record, as in 'This is going on your permanent record.' The web is now your permanent record. And everything is going on it."[82] These weak tie customer satisfaction systems are quickly becoming a way to quantify expertise and performance.

Measuring ties between nodes, whether from citations or recommendations, is far from the only technique for matching people to challenges. A related technique is *measuring credentials*. Credentials are how most organizations, companies, and institutions, including government agencies, measure who knows what. They are how most people conventionally think about expertise on a résumé—namely, as a designation awarded by a third party, such as an academic institution, an employer, or an association. Trademark law protects the ability of third parties to designate members as, for example, "Certified Public Accountant" with a collective mark. It also prevents anyone who fails to meet the standards set by a third-party organization from using a certification mark such as "Champagne," which is reserved only for growers with an appellation of origin in the French Champagne region. Credentialing is so important to finding a job and achieving greater professional success that three hundred million people from two hundred countries post their résumés on LinkedIn—with two new members joining every second.[83]

Traditionally, credentialing is a two-step process for measuring expertise. First, the reputation of the credentialing institution is weighed in the mind of the evaluator. Second, the qualifications are assessed against the reputation of the validating institution. The higher the reputation of the credentialing body in the relevant community of interest, the greater the value of the person's credential. Although a medical degree is a significant attestation to time and money invested, a medical degree from Harvard Medical School carries a different weight than a degree from the Institute for Functional Medicine. Harvard's top ranking in *U.S. News and World Report* makes a degree from its medical school a trusted predictor of expertise (and earnings potential) in some contexts. But among those who value alternative and holistic medicine, it might be negatively correlated with an assessment of expertise. Similarly, a degree from Yale Law School might equate to relevant expertise among academic law schools but not be worth much to Wall Street corporate law firms or those who disdain lawyers altogether.

Although much about credentialing processes have not changed, the Internet facilitates the process of recombining credentialing information in new ways and layering in credentialing information that previously might not have been easily available. Think back to the Health Tap example. Health Tap makes it much simpler to search for doctors with specific credentials, but also to weigh the source of the doctor's diploma together with publication data and more practical evidence, such as how helpful she is to patients who ask questions on the website. The credential becomes one of many hallmarks of expertise that the evaluator can decide how to weight. Degrees and academic awards can be gleaned from institutional bio-sketches, Facebook pages, research networking sites, and other sources and then combined in different ways. Where one website puts the primacy on the ranking of the institution, another might place more emphasis on the topic of the degree. Suddenly, the process of measuring credentials becomes more fluid. Not only is the Internet democratizing the process of awarding credentials, it is also democratizing the process of evaluating those credentials by controlling how they are displayed, weighted, and searched.

A third way to quantify expertise is to *measure experience,* especially successful achievements, including work completed, milestones reached, tasks and challenges undertaken, and time spent on an activity. Whereas techniques like pyramiding and accreditation are useful for finding the rare expert, the new tools for measuring expertise on the basis of work accomplished might be better predictors of useful participation and successful work. The top doctor, based on recommendations by peers or graduation from the highest-ranked medical school, might not possess the best bedside manner or be willing to answer questions via Health Tap. The ability to measure expertise based on experiences gives the searcher a wider array of choices to target.

Whether scientific or musical, some skills require hands-on learning and mastery. As the inventor Dean Kamen is known for saying, "You teach soccer by giving kids a ball, not having them spend twelve years learning soccer rules." New platforms are turning new kinds of experience—such as acts of good citizenship in an online community or the number of times I clicked or donated or responded to a call for help—into a quantifiable measure of expertise. On the Internet, we can now participate in an almost infinite variety of leaderboards that quantify and communicate our aggregate experience by tabulating actions taken.

GitHub is a place to collaborate on developing software. People are using GitHub as a way to showcase their experience, accomplishments, and productivity. More activity on GitHub signals a developer's skills and passions. GitHub provides online hosting for repositories of computer code, as well as tools for coordinating collaborative projects between multiple engineers and social networking features. A user's expertise can be tracked from the source code tree in a GitHub repository.[84] A user's GitHub profile features an auto-generated list of proficiencies, alongside a repository of projects created. These proficiencies are determined on the basis of activity—projects actively contributed to, activity history, and number of followers and interactions. GitHub profiles are being used to evaluate programmers' technical skills, as well as their ability to work well with others on complex tasks.[85]

Platforms that quantify actions taken, such as sites that enable one to find a yoga teacher or a math tutor, also measure the number of times a person is consulted. Research on online communities continues to demonstrate that past performance is a useful predictor of future performance.[86] The more people have participated, the more they will participate. Hence these metrics of actions taken could provide a useful way to match people to problems. Tony Hsieh, CEO of the billion-dollar online shoe and clothing company Zappos, says essentially the same thing: "I haven't looked at a résumé in years. I hire people based on their skills and whether or not they are going to fit our culture."[87]

When expertise is defined by experiences and demonstrated skills, it can also be quantified on the basis of challenges and money earned. A person can also demonstrate relational experience—the ability, say, to communicate about a field, which is a powerful indicator of expertise—by blogging or tweeting or answering questions on Health Tap or Stack Exchange or by participating in other knowledge-based crowdsourcing platforms. The Stack Exchange network consists of 124 community-based question-and-answer websites, as of this writing. Users accumulate points for participating in the community, which are used to unlock tools, features, and badges. Stack Overflow, the largest community in the Stack Exchange network, is a question and answer forum for computer programmers. In 2014, Stack Overflow had over 3.8 million registered users and more than 3.1 million questions asked with 4.5 million answers. According to their homepage, "Stack Overflow offers a careers site, Stack Overflow Careers 2.0. The careers site is targeted toward companies that are seeking top quality programmers. Careers allows Stack Overflow members to link their Stack Overflow profile with their online résumé (created on the site, or imported via LinkedIn), allowing employers to browse member contribution history to Stack Overflow."[88] Stack Overflow measures expertise based on the straightforward calculation of the number of questions a person answers and how others evaluate those answers.

There are numerous ways in which platforms and the social systems they reflect are quantifying and showcasing expertise—or what some have called "status markers" in the literature on online communities, online reputation, or social network analysis.[89] "Each platform," writes crowdsourcing researcher Natalia Levina, "makes countless design choices in promoting content and its producers. Should the platform display status markers associated with the content (number of likes), with the producer (number of followers), or both? How prominently should it display perceived worthiness of content? For example, YouTube prominently displays the number of times a video has been viewed, while Wikipedia subtly notes which articles are featured." Amazon uses an elaborate system of badges, but Twitter sticks to a few—number of followers and retweets.[90] Although the academic literature has helped to describe and categorize these different status markers and the underlying behaviors they draw on to measure expertise, it is only by experimenting that we can arrive at more definitive answers about what kinds of expertise measures are useful for public engagement. Online communities can provide insights about which Wikipedia contributors write the best articles and which people answer questions most effectively on a Q&A site. But there is much work to be done to advance the interrelationship between online status markers and offline institutions.

All of these developments are bringing the ability to find the most relevant expertise anywhere in the world, tailored to each individual user's specific requirements, ever closer to realization. But there are still significant challenges to doing for civic-related expertise what Google has done for documents: making sites accurate and searchable enough for institutions to rely on for smarter governance. "Expertise" was not a well-defined concept even before the transformation wrought by new technology, and the field is still quite new. Notwithstanding recent progress, technological obstacles remain.

Searching for and targeting specific expertise, even within a single organization such as a government agency, is not yet something we know how to do well or comprehensively or consistently. Much of the practical experience with expertise retrieval to date has been with the mining of knowledge within firms and organizations as part of knowledge management initiatives. For the most part, it was a practical topic for human resources departments to wrestle with in the context of their work on corporate intranets. With the development of next-generation web tools and large-scale interlinking of documents and data, there is now the potential to make a much broader range of expertise more universally searchable. Academic networking services such as VIVO and Profiles have pointed the way, but much less is known about the approach needed to match citizens to opportunities to do public service.

Although some studies have focused on identifying expertise on question-and-answer sites, for example, there is not yet well-developed computer or data science literature on more integrative, cross-platform approaches to expertise retrieval. Discussions of craft-learning, badging, and the like in the field of education have not yet cross-pollinated with expertise retrieval studies in web or information sciences. Those who focus on taxonomies for organizing expertise work primarily on academic networking. They are not in regular conversation with practitioners of corporate knowledge management, firms that do expertise retrieval in the financial and investment sector, web-based expert networking platforms in consumer domains, or challenge and crowdsourcing practitioners with deep knowledge of how to frame queries and describe the problem to be solved.

Far more cross-disciplinary research on expertise is needed if smarter governance is to take off. In the public sector, for example, the person most mentioned in any document is likely to be a press officer, not a subject matter expert—which means that standard document correlation techniques are not particularly useful for expertise retrieval in public contexts. A cultural and policy shift to require that all policy documents list relevant subject matter experts would help, as would conversations among lawyers and policymakers as well as web, data, and information scientists about how to overcome the many hurdles that remain.

In addition, some big ambiguities remain to be tackled. Today, searching for expertise largely means searching for people who are the authors of documents or are mentioned in them. But how far should we trust these names? Does the badge awarded to Ken Smith on Khan Academy belong to the same Ken Smith who wrote an article on SSRN? Does a Li Chen Twitter account belong to the same Li Chen who blogs on Medium? Are journal articles by Martha Boyer the person of the same name who answered ten questions on Stack Exchange and tutors on WyzAnt and Google Helpouts? To some extent, of course, authenticated badges help to solve this problem of author disambiguation (telling the difference between one Smith and another), but when combining badges with articles, books, tweets, and blog posts, the issue becomes harder and more urgent. The research and publishing community has had to confront the name disambiguation issue (as well as dataset disambiguation and article disambiguation) for some time. ORCID is a nonprofit standard-setting initiative to provide a persistent digital identifier to every author to ensure name integrity in connection with things like grants and journal article publishing. Just as the financial industry is undertaking efforts to disambiguate the names of companies in order to track risk in financial transactions, commercial platforms such as LinkedIn, Health Tap, Kaggle, LiveNinja, and Ideoba will have to do the same.

So, too, is there need to develop robust taxonomies to describe expertise in different domains with clear terms that are not duplicative. Automating the process of assigning tags to label people and their attributes may be fast, but it can come at the expense of clarity. Especially if the goal is to describe skills and experiences, not only credentials, creating effective schema will depend upon detailed input from those knowledgeable about their fields every bit as much as input from data, web, and information scientists, and librarians about meta-level organization. There are no standard definitions yet for the skills that equate to being a good plumber, capable cyber-security specialist, or black belt in data science. But the increase in the number of nanodegrees, microcredentials, and new opportunities to acquire knowledge independent of traditional college settings is likely to accelerate the push toward a "vocabulary" to describe these skills in order to make sense of the new accreditations.

The federal government's database of occupations (O*NET) describes the skills and tasks associated with a particular job, but this occupational data needs to be updated and made available in a machine-readable format so that new digital credentials can be correlated to the skills they imply.[91] Similarly, the creation of such skills taxonomies will, in turn, help to democratize the range of offerings because it will be easier to translate them into a description of skills that employers can understand.

Designing good interfaces that are easy to use and show the right amount of information is hard no matter what the topic or tool. It is especially difficult to know what information about a person is the right information to display. If, for example, a policymaker wants to search for someone with data science expertise to ask about governing with big data, he needs more than just a list of names. The system needs to tell him why it has suggested those names, without overloading him with information and without telling him too little.

There are no textbooks with solutions to these challenges. Careful, intentional experimentation is needed, followed up with feedback from both users and targets about whether a match was good, perhaps in the form of ratings and rankings of levels of satisfaction. Ongoing and active participation might be the best metric. But for such systems to evolve, improving the user interaction must be part of the process.

Learning how to do different searches better is still another key domain of inquiry. A decisionmaker will need to know precisely what he or she is looking for—for example, confidence that the person will participate constructively; an objective measure of expertise; evidence of skills; and/or formal credentials. Designing good processes is every bit as important as designing databases.

The Internet is transforming our ability to identify who knows what and target them to participate. When we do, new technologies are also making it possible for us to organize people into new forms of online, collaborative activity. A more diverse online community that fosters collaboration to generate actionable, useful knowledge is what scholars in the Russian Academy of Scientists have termed a "multinetwork."

Using the multinetwork, we can begin to imagine engaging more citizens in solving hard problems. Against the backdrop of a broader dissatisfaction with top-down models of learning, ivory-tower instruction, and insular governance, we would be exceedingly foolish not to make the most of our new, technology-enabled ability to democratize how we define, express, and measure mastery. Greater learning, more fluid labor markets, and above all, better, more legitimate governance depends on our doing so.

Experimenting with Smarter Governance

> With the growth of the American administrative state has come responsibility for policies, services, and decisions on which Americans' health, safety, and financial livelihood depend. Failures in bureaucratic expertise can produce devastating consequences.
>
> —Susan L. Moffitt, *Making Policy Public*, 2014

When the Air Force was struggling with the problem of pilots and civilians dying because of unusual soil and dirt conditions in Afghanistan, it did not know where to turn to get help quickly.[1] The soil was getting into the rotors of its Sikorsky UH-60 helicopters and obscuring the view of its pilots—creating what the military calls a "brownout." Reportedly, the policymaker tasked with addressing the problem in the Department of Defense (DOD) had no way to know that his colleague, the man sitting almost directly across from him, had nine years of experience flying Blackhawk helicopters in the field.

In the fall of 2008, the Air Force Research Lab (AFRL) piloted Aristotle, a first-of-its-kind-in-government expert search and discovery software platform, not dissimilar from some of the expert discovery tools discussed in Chapter 4. The software utilizes computer programs with associated algorithms that make it easier to sort and find people using criteria such as reputation, credentials, skills, and experience. In this case, Aristotle was aimed at preventing the recurring problem in large bureaucracies of not knowing who works there and what they know. The Department of Health and Human Services, for example, has literally thousands of different personnel databases to keep track of what some estimate—but no one is sure—to be ninety thousand employees. The hope was that Aristotle, by increasing awareness of both credentialed knowledge and relevant experience, could improve collaboration and problem solving across the DoD. The military described it as a "revolutionary tool to pull together multi-directorate expertise to tackle urgent warfighter needs."[2]

In practice, Aristotle was a searchable internal directory that algorithmically integrated people's credentials and experience from existing personnel systems and public databases, and from users themselves, thus making it easy to discover quickly who knew what and who had done what. Its designer, Danny Hillis, one of the fathers of parallel supercomputing at MIT and the former head of research for Disney Imagineering, had created Freebase, a semantic data storage infrastructure for the Internet that made it possible to organize personal profiles of people quickly and automatically from a variety of sources. (Google subsequently acquired Freebase, which is what powers the search engine's ability to give you a quick bio of George Clooney or a synopsis of a movie in the right-hand margin instead of just a list of links.)

For the military, the application of the Freebase project was called Aristotle, after Alexander the Great's personal tutor. "In those days, Aristotle knew pretty much everything there was to know," wrote Hillis in an essay in 2000.[3] "Even better, Aristotle understood the mind of Alexander. . . . Through Aristotle, Alexander had the knowledge of the world at his command." As Hillis subsequently summarized: "In retrospect the key idea in the 'Aristotle' essay was this: if humans could contribute their knowledge to a database that could be read by computers, then the computers could present that knowledge to humans in the time, place and format that would be most useful to them."[4] For someone whose career was made getting processors to work together, imagining massively parallel knowledge was only a small leap from imagining massively parallel computers.

Aristotle was an attempt to solve a challenge faced by every agency and organization: quickly locating credentialed, skill-based, and situational expertise to solve a problem. The promise was that Aristotle would make it possible to find expertise within the agency, where too often those with relevant operational experience were buried beneath layers of policymakers who pronounced on the issues but did not know how to solve them. Before Aristotle, the DOD had no coordinated mechanism for identifying expertise possessed by its employees.

As Dr. Alok Das, the senior scientist for design innovation who was tasked with implementing the system, explained, "We don't know what we know." By enabling greater use of in-house expertise, Aristotle was believed to be capable of saving government money over the long term. But budgetary constraints killed the project in 2013.[5]

Ultimately, the real disappointment of Aristotle lay not in its sunsetting but in the failure of government to recognize that targeting could improve the capacity of an organization. The project could have yielded powerful insights about the value of a system for collaboration and performance had the Defense Department only attempted to prove that it worked. At a more granular level, it might have shown whether searching for and targeting people based on their expertise, rather than simply posting an open call for engagement, does more to improve participatory and collaborative problem solving across a bureaucracy.

What really got lost was the opportunity to experiment—to learn precisely which approaches work better in which contexts, to which ends, and for which constituencies. Of course, at some level all government policymaking is an experiment in social engineering, but largely without a methodology or a plan for assessing performance. Technology-enabled platforms like Aristotle make it both possible and practically costless to measure data about how people use the tool. Digital engagement is susceptible to empirical research and experimentation in a way that "real space" engagement is not. Aristotle was a missed real-world opportunity to increase actionable knowledge and mastery. But when initiatives like Aristotle do not include experimental research designs from the outset, government can never learn enough to scale up innovations like graph-based searching effectively or create the underlying proof that more open and collaborative ways of working are more effective as well as more legitimate.

At each stage of problem solving, there is a need for specific, credible and relevant input, including facts, data points, opinions, and evaluations. But to create institutions that are better able to curate information and apply innovative thinking from outside, it is not enough to use the technologies of expertise to match people to opportunities to participate. Research and experimentation are also essential to drive learning about when and in which contexts smarter crowd-sourcing works. Fortunately, technology-based innovations lend themselves to the application of agile and empirical tests, because software can radically accelerate the speed of empirical observation, data collection, and putting learning into practice.

Keep in mind that Google comparatively tested—what's known as a/b testing—forty-one shades of blue to choose the one it would use as its hyperlink color.[6] Every user of the Microsoft Bing browser takes part in approximately fifteen different experiments per session to test different versions of the content and how it is used.[7] In a contro-versial project designed to study user behavior surreptitiously, Facebook tested the impact on subjects of subtle changes in the type of news (happy versus sad) delivered via its news feed.[8] The contro-versy stemmed from the fact that this study became public. Such be-havioral research takes place on Facebook all the time. With the introduction of more technology into governing, it is not a big leap to imagine a host of productive (and ethical) governance-related experiments—not just in the choice between policies, but in how gov-ernment makes policy in the first place. This chapter focuses on how to test the use of smarter crowdsourcing in the making of policy and why doing so is vital to the transformation of our institutions.

Testing Innovations in How We Make Policy

Whether left or right politically, governance practices often seem de-signed by "people hermetically sealed in the house of government" (as columnist David Brooks once described the budget process). We have policy walls in place that limit experimentation and diversity in poli-cymaking practices, and we make policy based on guesses rather than evidence (another reason we need expert networks).[9] Software entre-preneur and political commentator Jim Manzi writes in *Uncontrolled,* his paean to the importance of the scientific method, that we lack the opportunity in policymaking "to experiment and discover workable arrangements through an open-ended process of trial and error."[10] It is inexcusable, he laments, that we make macroeconomic decisions with global consequences with no means to know what works and why. He argues for the freedom to experiment with different regulatory rule sets at the most local level even when such experiments lead to the adoption of more coercive policies that limit personal freedoms in some places.

This Jeffersonian brand of what some describe as the New Federalism calls for a rollback of the centralizing regulatory policies of the New Deal in favor of states' rights. It is an intuitively appealing argument to test the downstream effects of different policy choices, such as whether lowering the speed limit or building roundabouts is more effective at reducing the number of accidents. In recent years, there has been renewed appetite in many Anglo-American jurisdictions for testing different policy interventions. David Halpern, a senior adviser to Prime Ministers Blair, Brown, and Cameron, fought to establish a Social and Behavioural Insights (or "Nudge") Unit to advise the government on the use of randomized controlled trials (RCT) in policymaking in Britain. Randomized trials—in use in medicine, business, and other fields to test effectiveness by comparing the outcomes of an intervention with a particular group against a randomly selected group that does not receive the intervention—are an excellent and inexpensive way to figure out what works.[11] In such a trial, the target population is divided at random into two groups with different interventions applied to each, such as two different techniques for delivering a service.

This standard practice among companies testing website designs is almost unknown in academic and government contexts. For example, Sage Bionetworks, a nonprofit biomedical research organization focused on openness and patient engagement, developed a revolutionary new electronic consent tool to enable iPhone users to sign up easily for clinical trials via the phone's app store.[12] But they could not assess its efficacy with comparative a/b testing of different designs because the institutional review board (IRB), the academic body that oversees and approves research in universities, deemed such experimentation to be unethical. In government there are no IRBs to disapprove, but the mindset of the professionals-know-best still hinders a culture of experimentation.

The historic resistance to experimentation in institutional decisionmaking comes from the tradition of closed-door and professionalized governance, but also from the practices of academic researchers themselves, many of whom are loath to label as "research" or "experiment" anything that deviates from the large sample sizes and neat conditions found in the lab. Even when researchers are focused on analyzing institutions—such as Nobel laureate Elinor Ostrom, in the work that she did over thirty years emanating from the Workshop on Political Theory and Policy Analysis at Indiana University—the conclusion is often the need to "isolate the experiment from external, confounding variables."[13] Precisely because we can never fully understand all the possible rule choices faced by a policymaker, let alone all their potential outcomes, she advocates analysis divorced from practice lest the waters get too muddied.

Sociologist Duncan Watts writes about creating a kind of jury pool of hundreds of thousands of online research subjects willing to participate in experiments. Economist Ted Castronova saw the potential for using virtual environments as large, persistent, living labs over a decade ago. In 2008, the MacArthur Foundation granted him $250,000 to develop *Arden: The World of Shakespeare,* a massively multiplayer online gaming research space where his team could document that people in fantasy games act in an economically normal way, purchasing less of a product when prices are higher, all other things being equal. Castronova believed that videogames or so-called *synthetic worlds* were an ideal way to understand behavior in the real one because it was possible, for example, to manipulate prices artificially and observe reactions.[14] Arden's sole purpose was research. Not surprisingly, the funders and the players abandoned the game and project, complaining both about the poor quality of the game play and the cost relative to the value of the research findings.

Top-flight social scientists like these "virtual labs," because they are controlled environments where the scientific method can be adhered to when conducting macrosociology experiments, including crowdsourcing experiments.[15] To them, online communities have the potential to provide larger sample sizes than they can typically access through offline research with human subjects. Still, the central actors and objects of their attention and inquiry are individuals studied in hermetically sealed environments. Although "computational social science" works with real-world data, experiments have generally not touched on the world of institutions.[16] They tend to focus on individual behavior and motivations (or groups and their proclivities), but not on the design or redesign of organizations.

Previously spurned by the academic and public sector as potentially reckless and unethical, testing inventions by replacing opinion with empirical facts using RCTs, when done well and with sensitivity for real-world outcomes, can help save taxpayer dollars and halt interventions that are wasteful or ineffective—even if such experimentation also calls into question the expertise of the public policy professionals.[17]

In their book *Nudge,* Cass Sunstein and Richard Thaler popularized the idea of bringing such experimentation into public practice and inspired the creation of Nudge Units in the Australian, U.S, and Singaporean governments (and the growth of the UK one).[18] Nudge Units study, for example, whether sending personalized text message reminders to people who have failed to pay court fines induces them to pay (and indeed it does) or whether a certain social welfare benefit has the desired impact. Comparing two identical groups, chosen at random, can control for a whole range of factors that enable us to understand, as between two large-scale interventions or smaller aspects of a single policy, what is working and what is not.[19]

Although there is lots of writing about improving the use of science in policymaking, Nudge work included, there is still little attention to the science of policymaking itself and only the whispers of an effort to test empirically new ways of deciding and of solving problems. Even if we vary where and how and at what level of government we restrict abortion or legalize prostitution or change the design of the forms used to enroll people in social programs, we do not often question underlying assumptions about how we make these policies in the first place. Justice Louis Brandeis's "laboratories of democracy" might cook up different policy prescriptions, but they all use the same institutional beakers and burners they have always used.

Examining the choice between policies arguably helps to spur a culture of experimentation, but it does not automatically lead to an assessment of the ways in which we actually make policy. To be clear, the Nudge Units often test choices between policies, or sometimes test the impact of changing default rules, such as how and when to enroll people in a social service. Typically, however, they do not experiment with approaches to *making policy*. They do not examine how governance innovations, such as open data, prize-backed challenges, crowdsourcing, or expert networking, translate into results in real people's lives. Does open data lead to greater accountability by health care providers or more jobs for health care companies? Does crowdlaw lead to legislation that is more legitimate? Do prizes create enough incentive to develop measurably better solutions to the challenges of poverty? Does crowdsourcing how we tackle gaps in educational outcomes improve reading scores? And does identifying and targeting who-knows-what, such as when the DoD uses Aristotle, translate into more practical engagement and, ultimately, better outcomes in decisionmaking? These are not the questions they ask.

The first step toward implementing smarter governance, therefore, is to develop an agenda for *research and experimentation*— whereby we test the use of targeted crowdsourcing at the different stages of decisionmaking, including identifying problems, prioritizing an agenda, identifying solutions, choosing which solutions to implement, and assessing what works—to prove when and how it makes sense to use outside expertise. We need an empirical research agenda for studying the impact of such smarter crowdsourcing on how policy gets made and how problems are solved.

Creating such research agendas is exactly what the Tobin Project does. In its mission and work plan, the Tobin Project in applied social science research, which is based at the Harvard Business School, follows the example of Dr. Judah Folkman (1933–2008), a noted cancer researcher. Folkman kept his lab focused on "big ideas and challenges, amidst the skepticism and the pressures to take smaller, surer steps," with a whiteboard of unanswered questions. As explained in a Harvard Business School case on the lab, Folkman had long "maintained a list of questions he deemed fundamental to understanding angiogenesis on the conference room's whiteboard. . . . Some questions stayed on the board for only a few weeks; others remained for years, waiting for new technology or a young scientist who would accept the risk."[20] At my own organization, the Governance Lab, we maintain a virtual version of the whiteboard called "If Only We Knew . . ." that we use to develop the list of important unanswered research questions on the impact of governance innovations.[21]

Imagine if such a research agenda had existed before the adoption of Aristotle. It would have been obvious to use an RCT to divide the target audience at the DoD into two groups. One group would have had their information automatically uploaded and their profiles prepopulated; the other half would have received a blank form to fill out. Observing the differences in outcomes would have yielded real insight into tactics for improving how the DoD works. Imagine if some people had been asked to submit their credentials and others to say something about their skills. Who would have been more likely to participate? And who, when later contacted with a request, would have been more likely to answer? Armed with such evidence, further investment could have been justified. We'll never know.

Today, we are witnessing the birth of new disciplines concerned with how people can become more intelligent by combining human and machine intelligence to organize knowledge.[22] Even with these technological advances, there is still surprisingly little *research* about how complex public organizations use information to make decisions and solve problems. Although resistance to experimentation comes, in part, from those who govern, social science researchers are also reluctant to label explorations of, say, the impact of crowdsourcing or expert discovery on downstream outcomes as formal experiments because there is no gold standard of scientific practice for conducting real-world experiments.[23] Hence it was not until early 2015 that the newly constituted World Bank Digital Engagement Evaluation Team began revealing the results of pathbreaking academic studies of datasets from online citizen engagement sites. They found, for example, that people who said "please" when they complained about potholes or broken red lights via citizen complaint websites were more likely to get a response and that men were dramatically more likely to get a response than women.[24]

There is already a great deal of scholarship about what motivates people to join a commercial crowdsourcing exercise, how they participate, and which features enable sustained and successful collaboration. There is also a great deal of attention paid by think tanks to the debate about choice of policy. But there is next to no research on the practices institutions use to collaborate with citizens to solve problems that touch people's lives. There is almost no empirical testing of alternatives in real-world contexts. White House innovation expert Tom Kalil is fond of saying that the Department of Education gives away two-thirds of its budget in grants and loans for education, but absolutely none of that money goes to experiment with learning how to create a Department of Education that works better. If we are to deepen our understanding of how institutions and organizations can best use specific expertise in specific contexts, we urgently need empirical research and experiments that "hover low over the facts."[25]

What Aristotle failed to do in 2010, the Food and Drug Administration (FDA) began to do in 2014. The FDA introduced expert networking tools akin to Aristotle to find needed expertise within the agency and won a $350,000 internal government competition in 2015 to expand their use across the whole FDA, and subsequently the Department of Health and Human Services, to improve the regulatory review of medical devices. The FDA has also embraced the use of experimentation—testing alternatives—as it develops this new program, called Experts.gov.

As of this writing, the FDA is ready to begin staffing some medical device review panels using the new software. Other device review panels will be constituted through the traditional process of calling around. The goal of these experiments—which the Governance Lab designed and planned in collaboration with scholars in the MacArthur Research Network on Opening Governance—is to yield real-time results that demonstrate whether and how targeting participation improves the outcomes of regulatory review.

The FDA's willingness to start with a research agenda and test and explore the impact of changes in how it operates provides a blueprint for how to go about adopting new technology-based innovations for governing. Real-time experimentation is crucial. All governments, whether in Brussels or Brasilia, despite plummeting rates of trust, resist changing how they work. It is, after all, hard to reengineer the plane while flying it. Although practices of open innovation are as radical a change as they are promising, shifting from "closed source" to "open source" government will not, and frankly should not, happen without evidence that open source works. Companies demand proof that shifting from closed to open source software generates a return on investment before they commit their time and money. Government should, too. Agile and empirical experimentation, conducted in parallel to the rollout of a new governance innovation, can provide that evidence. And it is eminently doable when the innovation is delivered via a configurable and changeable software platform. Without experimentation, there is no way to know whether the change is worth the effort.

The ExpertNet Consultation

Shortly after Aristotle died, the Obama administration made a public commitment as part of its National Open Government Action Plan in 2011, and again in 2013, to expert networking (though not yet to experimentation) that laid the groundwork for current efforts.[26] Earlier, the president had called for engaging the knowledge, expertise, and perspectives of diverse members of the public, saying, "Knowledge is widely dispersed in society, and public officials benefit from having access to that dispersed knowledge," and hence to "collective expertise and wisdom."[27]

The White House endeavored to make good on the president's promise by launching an online discussion via a wiki about the desired features of new engagement platforms. This consultation was called the ExpertNet project (and was the forerunner to Experts.gov).[28] The wiki solicited input about platforms and processes for targeting and matching public expertise to opportunities to participate. The proposal laid out a number of potential uses for which such expert engagement could be valuable. Here is one fictional use case: "The Department of Innovation wants informed, public input on the department goal of improving the literacy and mathematical skills of adults seeking advancement."

Then the consultation listed questions that this fictitious agency would want to ask experts, such as: What empirical evidence exists about the relationship between adult academic performance/career advancement and incentives? What effect does the structure of the incentive—whether explicit or implicit, predictable or unpredictable, tangible or intangible—have on outcomes? What are the methods of identifying incentives that adult learners respond to? What criteria might influence the incentive effect (e.g., ethnic and family wealth differences) or the attractiveness of different incentives?[29]

The ExpertNet consultation clarified that expertise is defined broadly and that questions could be targeted to those with relevant skills—such as those who work in private and public sector adult education, active adult learners, and those with expertise in reward-based academic performance/career advancement programs. In addition, it found that the mythical Department of Innovation could benefit from the advice of experts in fields such as data analysis for creating visualizations to track academic gains, effective software programs, and games that effectively teach adults specific skills.[30]

This proposal (which I wrote) sought public input about the platforms, techniques, and processes available for discovering expertise, including tools designed to do expertise searching and to process needed refinements, to find examples of others using expert discovery, and to discover any legal or other impediments. The conversation, which was highly detailed, attracted advice from forty-two registered users, including professors, industry experts, and activists. The goal was to get their advice about running pilot projects to test the suggested strategies using different platforms.

The consultation, however, did not result in any immediate implementation of ExpertNet at the time because we could not find a senior government official willing to ask questions of the public that might reveal that he did not have all the answers. Despite the absence of any forum for targeting those with specific know-how, the project was declared over with the signing of a free contract by the federal government with the Quora Q&A website. But there were no plans on the part of federal institutions to make use of it or any other tool to improve governance.

Experimentation at the FDA

Now, several years later, the Food and Drug Administration is launching its own implementation of ExpertNet called Experts.gov and starting to fulfill the promise of these early efforts. This project goes beyond hypotheticals to experiment with *targeting* expertise in actual practice. It uses a modified version of a platform for cataloging and searching for academic biomedical experts. As an agency that depends heavily on scientific and academic expertise, the FDA is a natural first mover for targeting participation by those with specific forms of know-how. (It is worth noting that the White House had to revoke its official support of this effort lest it run afoul of the Federal Advisory Committee Act, the statute that governs how the federal government gets expertise. Chapter 6 discusses in some detail how this and other legislation often impede such efforts.)[31]

The FDA regulates $2 trillion worth of products every year from food to drugs. Among its important responsibilities, the FDA is tasked with reviewing the efficacy and safety of new medical devices—from hip implants to heart stents—that reduce suffering, extend lives, treat diseases, and generate enormous economic gain for successful inventors.[32] The market for medical devices is estimated to be $140 billion a year, and spending on medical devices is growing at a rate of 6 percent—more than the 4.5 percent growth in health care expenditures generally.[33]

The Center for Devices and Radiological Health (CDRH), a division of the FDA, manages this complex process of premarket approval and postmarket review of the most dangerous medical devices in response to submissions by applicants, including clinical trial data. CDRH's Office of Science and Engineering Laboratories (OSEL), the research hub of CDRH, employs its own scientists and also consults with outside experts, whose job it is to help the agency understand the market and be in a position to assess the safety and effectiveness of medical devices from invention to eventual use.

The FDA separates medical devices into three classes, each meriting a different standard and process of review. Class I are the lowest-risk devices, such as tongue depressors. Class II are intermediate-risk devices like contact lens solution. The smallest category, Class III includes the highest-risk, usually implantable, devices, such as pacemakers. Although these high-risk items represent only about 1 percent of the devices that the FDA regulates each year, they account for an outsized fraction of medical device spending. In 2008, spending on the six highest-cost implantable devices alone was about $13 billion—or approximately 10 percent of total U.S. medical device spending.

Although the pathway to compliance for low-risk items like Band-Aids is straightforward because they are exempt from regulatory review; more complex, life-sustaining, high-risk devices such as breast implants require judicious premarket approval by those with the right know-how, including engineers and experts in manufacturing quality, biocompatibility, and epidemiology. Just imagine the diversity of know how required to opine on the safety of an edible battery versus a heart monitoring app or a 3D-printed exoskeleton. If medical devices reach patients before being properly tested, the consequences can be dire. Dr. Steven Nissen of the Cleveland Clinic estimates that faulty devices contributed to more than 2,800 deaths in 2006, for example.[34] In a CBS report on medical implants, Dr. Nissen affirmed, "People make the assumption that when their doctor implants a device, whether it be an artificial joint or a pacemaker, that it's undergone very rigorous testing. That assumption isn't always true."[35] From breast implants to metal hip replacements, there have been recalls affecting thousands of people.[36] In only 21 of 113 cases of recalled devices between 2005 and 2009 had the agency conducted a thorough premarket review. Put another way, the agency had exempted most of these devices from review and imposed less stringent review processes when a device resembled something that was already on the market.[37]

But delays in approval can also cost lives and dollars. In one extreme instance, Acorn Cardiovascular developed a device that helps to shrink an enlarged heart, but it could not market the device because the FDA demanded an additional three-year clinical trial, which cost the company $30 million. And the medical device company Biosensors International shut down its operations in California owing to the time and expense associated with getting FDA approval for a cardiac stent—a device already available globally, including in Mexico and Canada.[38]

Harvard Business School Professor Ariel Stern, who studies the economics of medical device review, has found that "pioneer entrants spend approximately 34 percent (7.2 months) longer in the approval process than the first follow-on innovator," potentially depressing innovation. It is routine procedure for an applicant to go through at least two rounds of back-and-forth review before a final determination by the agency. Stern has also found that a one-month increase in the length of the regulatory approval process is associated with a 1.6 percent reduction in the probability of any subsequent adverse events and a roughly 1.1 percent reduction in the probability of a subsequent adverse event involving patient death or injury. Clearly, balancing safety with alacrity is not an exact science.[39] It is also necessary to contend with political pressures from both business and consumer groups. Although the FDA does not issue clear data on backlogs or processing times, research points to shortcomings within its current system for researching and approving complex devices.

Deficits in Regulatory Review

In the FDA's current model of premarket review, it can be difficult to find the right expertise to staff a review team, which typically comprises ten to fifteen members.[40] Shortages of *agility, know-how,* and *diversity* are problems shared by many organizations in a variety of domains.

Premarket review and approval are designed to protect consumers by ensuring that devices are safe and effective, but also to accelerate the path to market for innovative and potentially life-saving and lucrative tools, At present, it can take nine months simply to find and convene a qualified review panel from the ranks of agency employees. A panel typically includes one or two medical officers, one statistician, one epidemiologist, one or two biocompatibility experts, one packaging and sterility expert, and three or four engineers, depending on the product, plus a manufacturing expert and a senior manager.

The FDA has long been criticized for being too slow in approving medical innovations. Agility is one of the biggest challenges, especially in light of the interdisciplinary skills required for review (not unlike mine, workplace, consumer product, nuclear, and other government safety reviews). With the advent of new kinds of devices, including "more combinatorial products, more large molecules, nano-delivery systems, and many more innovative approaches" there is an ever-more urgent need for cross-disciplinary expertise.[41]

Only about one hundred scientists work at OSEL (eight hundred total at CDRH), and more expertise is always needed in order to monitor the safety of devices throughout their lifecycle. From 2000 to 2010, the average time to decision on high-risk but potentially life-saving devices increased by nearly 60 percent (from 96 to 153 days). Over approximately the same time period (2000 through 2009), the number of submissions for premarket approval remained the same, hovering just below 10,000 (9,594 in 2000 and 9,655 in 2009).[42] To help speed things up, the FDA has introduced new information technology (IT) decision support tools and a policy of giving clearer guidance about procedural steps, including key scientific and regulatory requirements, that need to be satisfied within a 120-day period. But in order to map out a more predictable yet customized regulatory pathway, the FDA still needs much more expertise faster, including more internal expertise. Because of budget cuts, it cannot simply hire people.[43]

In January 2011, the FDA committed to a plan of action, known as the Innovation Pathway, to transform its regulatory process and improve how it works with entrepreneurs both to bring new devices to market and to protect the public.[44] The Innovation Pathway pinpointed the expertise deficit as a key problem, citing the need for greater collaboration with a larger network of experts.[45] Done well, the use and re-use of the same pool of people to conduct reviews creates a high degree of institutional knowledge. But that is not always what happens. "When you receive a letter back from the agency in response to a submission," writes one commentator, "you are potentially dealing with a very bright, well-educated reviewer, who may have very little or no industry experience, and may have little agency experience due to turnover and attrition."[46] Likewise in other regulatory agencies, such as the Patent Office, a patent examiner may not have any training in the specific science or technology at issue in a patent application. Nonetheless, to avoid self-dealing and conflicts of interest, many agencies have statutory prohibitions against soliciting outside input.

CDRH uses outside experts for advisory panels, but membership on a device review panel is limited to a pool of Special Government Employees who must be recruited and enrolled.[47] The Center cannot be assured that an expert in an emerging technology will be available exactly when that expertise is needed. Other traditional ways to identify external expertise, such as public workshops, conferences, and literature reviews, may lag behind current research or may not be available when a scientific question arises at CDRH.[48] Outside experts are most often used just to increase the state of knowledge in general but not to participate in device reviews.

This lack of expertise results from high reviewer and manager turnover at CDRH (almost double that of FDA's drug and biologics centers); insufficient staff training; extremely high ratios of frontline supervisors to employees; and CDRH's rapidly growing workload, caused by the increasing complexity of the large number of submissions.

The Innovation Pathway originally called for a pilot to create a vetted list of experts across two dozen membership organizations such as the American Academy of Neurology and the Society of Thoracic Surgeons to serve as advisers and reviewers.[49] It relied on membership in these professional associations to serve as a proxy for expertise without any means to "match" people to problems with any specificity. Although membership in the Society of Thoracic Surgeons might, for example, require that one be a practicing thoracic surgeon, it says little about prior experience with different technologies and devices, skill as an adviser or reviewer, or interest in the kinds of problems members would like to work on. Above all, the vetted list does a poor job of reaching all those with relevant experience—a PhD candidate at a university, say, whose dissertation is on a specific medical device, or an academic or physician who does not belong to one of those associations.

As a result of the Innovation Pathway and some other measures, the FDA's backlog began to drop significantly in 2010; but the gains are likely to be short-lived.[50] A study completed by the Boston Consulting Group points to a significant rise in the number of drugs and complex devices approved in the European Union long before being reviewed in the United States. As the devices get more complex, the United States is likely to fall even further behind.[51]

In addition to being shorthanded, the device review process also falls short in its diversity of expertise. Even for a common device, the FDA has to find more than a dozen qualified people in a variety of specialties. But what about finding people knowledgeable about cutting-edge innovations in new fields? Devices combine many different fields of endeavor from materials science to fluid mechanics. Imagine trying to convene the appropriate reviewers for a wearable or implantable device that is made of a new kind of adhesive material, incorporates wireless transmitters, and delivers a biologic substance to treat a condition. Also, because new devices rely on new kinds of science, they are sometimes hard to describe and are often called by different names in different fields.

The proliferation of mobile health devices such as heart monitors that leverage the sensors in a cell phone has added further complications. The Federal Communications Commission (FCC) regulates mobile devices; the FDA regulates medical devices.[52] The complexity of having to clear hurdles in two agencies led the federal government to issue a memorandum of Understanding in July 2010, promising to "improve the efficiency of the regulatory processes applicable to wireless-enabled medical devises."[53] But many feel progress has been too slow. In April 2012, six U.S. representatives sent both the FDA and the FCC (at the behest of the mobile technology industry) a letter urging the agencies to improve communication so as to avoid issuing "slow and inconsistent regulation" that delays the progress of wireless medical technology.[54]

The 2012 Food and Drug Administration Safety Innovation Act (FDASIA) included a requirement that the FDA, FCC, and Office of the National Coordinator in the White House work together to publish a "proposed strategy and recommendations on an appropriate, risk-based regulatory framework pertaining to health information technology, including mobile medical applications, that promotes innovation, protects patient safety, and avoids regulatory duplication."[55]

The agency is also responsible for keeping tabs on device safety from cradle to grave. It needs to know about a diversity of device types and industries, as well as economic, clinical, and policy impact. These responsibilities also require different types of know-how. In addition, there is always the need for insight from people in fields where one would not automatically assume any relevant prior knowledge. For example, an entomologist who studies the architecture of consistently cool anthills in the African savannah may know a huge amount about green building design.

Improving Device Review with Targeting

Previously, the FDA had wrestled with the idea of allowing agency outsiders to participate in reviewing medical devices, but potential conflicts of interest and accusations of undue influence led them to abandon the idea. So the agency continues to rely upon its own internal staff augmented by some outside experts, who are hired as temporary employees to staff the review panels. It uses outside experts primarily to provide background information rather than actual device review.[56]

New expertise platforms have the potential to reduce the cost of targeting the expertise the FDA already has but cannot find quickly within its own walls. They offer better ways to identify expertise by means other than just university diplomas; to search for and find relevant experts across multiple disciplines beyond those known to the searcher; and to match expertise, including experience in the review of similar devices, to a hard problem, such as the steps involved in regulatory practices with real-world consequences.

Targeting expertise, of course, should not rule out serendipitously finding the right people through an open call in a large, unknown pool of participants. Accordingly, the FDA's project starts from the modest supposition that a directory of skills and expertise can be of use within a known population to match people to the right opportunities to participate. In the case of medical device review, however, the need for quick access to reliable expertise far outweighs the unreliable benefits of serendipity, especially when legal and cultural practices, as well as the very real fear of industry manipulation and abuse, make it unlikely, even impossible, for standard crowdsourcing practices to succeed.

Sourcing expertise in highly technical fields, where it can often be found among academics and regulators who maintain careful data about their qualifications, is a natural first foray into expert networking. Academics reliably maintain records of their credentials and publications. The norms of the profession require sharing such information openly, and databases are already prevalent in the biomedical sciences more so than in other domains. Regulators are less transparent about their credentials but more so about their experience, at least internally. There is a record of which regulator worked on which regulatory action. In short, there are datasets for these two groups that could aid in populating profiles and automating requests to find those with specific know-how.

There is no shortage of knowledgeable people in the biomedical sciences. In addition to the 800 scientists at CDRH, approximately fourteen thousand people work at the wider FDA, another seventeen thousand at NIH, over eighty-three thousand at their parent agency,

Health and Human Services, and hundreds of thousands in industry and academia.[57] Although the FDA's Medical Devices Advisory Committee comprises eighteen panels of medical practitioners and industry experts representing consumer and industry interests, the agency has no way to identify quickly enough diverse individuals internally with the relevant expertise to review specific applications.[58] Neither can it find knowledgeable participants in the private sector who might have deep knowledge of an area innovation. On occasion, relevant portions of a premarket review dossier can be sent to an advisory committee to pose questions and get additional information. Obviously, what the FDA needs is a consistent means to find both external and internal experts in connection with regulatory review panels, as well as to educate the agency about new trends in medical devices, such as identifying the therapeutic benefits of 3D printing or nanotech.

Enter FDA Profiles. The Office of Science and Engineering Laboratories at CDRH launched a "Research Networking" software pilot (not unlike Aristotle) called FDA Profiles. The project got started under the leadership of one of the FDA's new "Entrepreneurs in Residence," outside innovation fellows invited to inject innovative ideas and insights into government and educate those in industry and academia about an agency's priorities and challenges. These fellows, one of whom was the head of the Health Information and Decision Systems Program at the University of Maryland's Business School, are given the mission of helping CDRH transform and accelerate the device approval process.

Profiles was not an attempt to upset the applecart of power. The purpose of using software to match people to the opportunity to participate was not to change who makes the ultimate decision at the FDA about devices. On the contrary, the project's ultimate goal was to improve the current regulatory process for device review by increasing both efficiency and quality. Profiles uses the Harvard Profiles Research Networking Software platform developed by Harvard Medical School with support from the National Institutes of Health; it is in use at 240 institutions, including the medical and biomedical faculties at Harvard, Pennsylvania State University, Boston University, and the University of California at San Francisco (UCSF).[59] Profiles is an open source expert discovery and networking tool funded by the NIH, whose primary use is to aid researchers in finding colleagues with specific areas of expertise. By one description, "it imports and analyzes 'white pages' information, publications, and other data sources to create and maintain a complete searchable library of web-based electronic CV's."[60]

The FDA has begun adapting Profiles to create a searchable online expert directory, initially of FDA staff and then of scientists within Health and Human Services, using both internal FDA information such as personnel data and external sources such as publications from PubMed and Web of Science. The goal of the first phase of the project, begun in 2014, is to enhance the speed and efficacy of medical device review by using networking tools to facilitate collaboration between internal FDA scientists and regulators and, eventually, to enhance collaboration between government regulators and academia in understanding the complex science needed to make hard policy decisions.[61]

How Profiles Measures Expertise

Harvard Catalyst Profiles is a classic example of an expert network. It is a good example of how a future of smarter governance could work. Profiles measures expertise through a combination of passive and active means. Through so-called passive networking, its algorithms are designed to add publications automatically to a person's Profiles bio from online sources and then analyze these publication data to build a catalog of a person's expertise and research network.[62] Passive networks are automatically created based on current or past co-authorship history, organizational relationships, and geographic proximity. Profiles scrapes from online biographies, professional society memberships, education and training records, affiliations, regulatory accomplishments, publications pulled from open databases, and grants culled from grants.gov to autopopulate a profile.

In addition, a researcher can manually add data about his or her interests, skills, and projects or specify his or her adviser, mentor, or relationships with other people, "expand[ing] . . . [Profiles] content with information about social networks that only they know." With the combination of passively gleaned and actively contributed data, Harvard Catalyst Profiles speeds the process of locating the "right" experts by algorithmically matching people to opportunities to participate.

The organizing framework for all these data—whether scraped from databases or manually inputted—is borrowed from VIVO, another open source expert discovery initiative. VIVO is a multi-university collaboration to synthesize authoritative data about researchers.[63] "When installed and populated with researcher interests, activities, and accomplishments, [VIVO] enables the discovery of research and scholarship across disciplines at that institution and beyond." VIVO supports browsing and a search function for rapid retrieval of desired information.[64]

VIVO's working groups of librarians and information scientists have developed an ontology—on which Harvard and FDA Profiles rely—for describing a biomedical researcher and his or her expertise.[65] This ontology provides the organizing categories and common definitions for describing research fields, research projects, datasets used, teaching experience, and of course fields of expertise. The Profiles tools use the VIVO ontology to support both passive networking (culling from citation networks—that is, who cited whom, who co-authored with whom) and active networking (manually entry of data about people's projects and interests). From both these sources, Profiles generates a set of prioritized keywords—what they call the Concept Cloud—to define a researcher's interests over time.[66] People can then add their own skills and edit their own profiles.

The Profiles platform uses an algorithm to return relevant search results in a list and in visual graphs showing relationships between people. If you type in "stent" and FDA Profiles, for example, you will get a list of people at the FDA along with explanations of why they are suggested. Although Profiles initially functioned as a kind of "FDA Facebook," the plan has been to expand its use more broadly within government and interconnect with other networks like VIVO to find expertise outside the agency. In the future, there are also plans to incorporate into it patent data as well as data from LinkedIn, Slideshare, and other social media. And there are experiments anticipated to integrate discussion and collaboration software with its people-search capacities. Once one finds relevant experts, it is important to be able to communicate and collaborate with them. UCSF, for example, is experimenting with Open Proposals, a crowdsourcing site to engender precompetitive collaboration among academics, which is already generating robust and productive engagement.[67]

Smarter Governance in Practice: The Experts.gov Experiment

The rollout of FDA Profiles dubbed Experts.gov created an opportunity—the opportunity squandered with Aristotle—to go beyond merely deploying new software and test what works through empirical, agile, pilot experimentation. The specific goal of such experiments is to assess the impact of expert networking on the safety and effectiveness of the FDA's premarket and postmarket regulatory reviews of medical devices. These experiments will tell us more than simply whether such targeting works. They will also tell us whether it helps to increase intelligence in the system so that institutions get smarter.

FDA Profiles affords a unique chance to study the impact of expert networking on crowdsourcing generally—and in the context of regulating product safety, more specifically. Perhaps more important, it makes possible a first-of-its-kind, real-time experiment with how government makes decisions. Comparing what happens when a general call to advise or participate in a regulatory review panel with what happens when participation is sought using Profiles will advance our understanding of how to introduce approaches to policy-making that are empirically validated and agile in design.

It is worth delving into the details of these experiments—designed and planned but, as of this writing, yet to be executed—if only to illustrate how such technology-backed innovations in smarter governance lend themselves to the testing and experimentation that can and, in fact must, accompany their introduction. Adding empirical research to FDA Profiles is a chance to experiment not simply between policies, but with how we make policy.

So, for example, even within the target population of FDA and Health and Human Services personnel that Profiles is uploading into its database, it is possible to compare targeting participation with Profiles against status quo methods to understand the impact both of general matching and of matching based on academic degrees and publications. Does targeting on the basis of academic credentials have the downside of reaching those who are too busy to help and whose professional norms discourage such volunteer activity? Good academics may, for example, turn out to be bad participants because they are trained to contribute in ways that are longwinded and unhelpful to regulators.

After comparing the use of Profiles against its absence, the plan is for experimenters to run parallel trials with tweaks to the algorithm in order to weigh some forms of expertise more heavily than others. In order to test the impact of targeting, the FDA plans, in some instances, to sort people by academic degree and publication record; in others, to redefine expertise based on experience reviewing devices; and in still others, to scrape data about them from sources such as LinkedIn, Facebook, and Wikipedia. Finally, data from third-party sources can be combined with asking people to complete a questionnaire to self-report skills and experience. Testing a variety of criteria for selection is now part of the rollout plan.

There is at least one direct precedent for such experimentation "in the wild" with expert networking, albeit not in government, in the experiments run by Harvard Business School professor Karim Lakhani and his colleagues. Lakhani, an expert in open innovation practices, studied the impact of open innovation in the Harvard Catalyst project on diabetes research. Harvard Catalyst (the parent organization that created Profiles) is a "pan university clinical translational science center" located at Harvard Medical School that explores how to rethink all aspects of the scientific research process to help "high-risk, high impact problems in human health-related research."[68] Catalyst takes each aspect of the research process and designs a strategy for opening it up to more information and input from a wider audience. The goal is to bring in fresh ideas and novel perspectives.

In 2010, Catalyst undertook an experiment designed by Lakhani to open up how universities generate research questions.[69] Typically, an academic decides on the direction for his or her lab. In an effort to generate new ideas from unlikely sources for promising approaches to fighting Type 1 diabetes *before* investing research funding, Catalyst sponsored a $30,000 prize-backed challenge to come up with promising research topics. The Catalyst challenge was unique because it did not ask people for answers; instead, contributors supplied the questions. This format enabled people to suggest ideas whether or not they had the resources to solve the problem.

After six weeks, 150 solid research hypotheses had been submitted, encompassing a broad range of approaches from different disciplines. The authors analyzed the subject matter of the submissions and found that they were "quite different from what existed in the literature and from the existing body of ideas under investigation within the Type 1 Diabetes research community."[70] The Leona Helmsley Trust put up $1 million in grant funding at Harvard to encourage scientists to create experiments based on these newly generated research questions.

In addition to normal advertising of the grant opportunity, Harvard Catalyst used Profiles, the same technology underlying FDA Profiles, to identify researchers whose record indicated that they might be particularly well suited to submit proposals. The Profiles system takes the PubMed-listed publications for all Harvard Medical School faculties and creates a database of expertise based on a classification of their published papers. The topics of the experimental design proposals were matched to classifications within Profiles to facilitate searching. "The intention was to move beyond the established diabetes research community and discover researchers who had done work related to specific themes present in the new research hypotheses but not necessarily in diabetes."[71]

In the end, the matching algorithm yielded more than a thousand scientists who potentially had the knowledge needed to create research proposals for these new hypotheses. Some were accomplished diabetes researchers, but many others were not. Harvard Catalyst then emailed these faculty members and announced the new funding opportunity. The team introduced a control at this point by randomly selecting half of the targeted researchers to suggest additional people as potential collaborators, identified by Profiles as being complementary, and suggested that they work together. As in the public sector, this kind of targeted outreach is not common in biomedical research.

The outreach resulted in thirty-one Harvard faculty-led teams competing for funding, of which twenty-three had been identified using Profiles. Of those, fourteen had no significant prior involvement in Type 1 diabetes research. In the end, seven proposals were funded; five of the lead investigators had not done any prior work on Type 1 diabetes. It is too soon to know the impact of this experiment on sufferers of the disease. The typical timeline for medical innovations translating from bench to bedside is fifteen to thirty years. But the experiment did yield some exciting results: "The core insight driving Harvard Catalyst's experiments was that all stages of the previously narrow and fully integrated innovation system—from hypothesis generation to idea selection to execution—can be disaggregated, separated and opened to outside input. By opening up participation to nontraditional actors, Harvard Catalyst achieved its objectives of bringing truly novel perspectives, ideas and people into an established area of research."[72]

This kind of "wild" experimentation shows that there is absolutely no reason the regulatory process for examining devices should not be subject to the same methods as the process for testing the devices themselves. Doing so would pave the way for research on: (1) identifying and matching expertise; (2) defining and framing the call; (3) creating incentives to participate; and (4) developing institutional readiness. But remember, empirical research here is not an end in itself. It is a means for making better decisions about investments in new technology, understanding their impact, and improving regulatory review—that is, making governance smarter.

The Research Agenda for Smarter Governance: Identifying Who Participates

FDA Profiles has its own algorithms for identifying potential experts within and outside the FDA. By mining grants, publications, and bio-sketches, the Profiles approach draws from significantly different data sources than, for example, an expert network comprising members of accredited organizations. This raises a core question in identifying expertise: Which approaches work best to produce both willing and high-quality participants? There is no guarantee that targeting based on credentials will yield the "right" community of participants. In fact, the whole problem is that we do not know up front who and what we do not know. The appeal of *crowdsourcing* and open innovation is that these practices will attract unexpected participants. In the Harvard Catalyst case, the search was broad, resulting in a thousand potential candidates, many of whom had no direct training in diabetes research. But the hope is that *targeted crowdsourcing* will make it possible to find the "right" people with more certainty and regularity. It is for this reason that multiple government agencies both in the United States and abroad are beginning to explore the adoption of expert networking tools to inform how they govern.

Empirical experiments can provide answers to important questions, such as:

- How do expert systems identify who knows what?
- How does a platform target requests to participate and match people to opportunities?
- Who participates more readily and more usefully? Those whom Profiles define as experts or those whom the agency finds via LinkedIn?
- How useful are these techniques for uncovering diverse experts? In other words, it may be simple to find people who work on heart stents, but what's the best way to find a materials scientist who works in another domain but could have relevant insights to share?
- Are these techniques more useful for staffing review panels or getting insights about next-generation devices or other policymaking activities?

How the call to participate is described is also configurable and testable (by using different language to articulate the problem) and subject to trial and error. Asking people to participate on the basis of the urgency of the problem might be compared with asking them to participate on the basis of an appeal to their interests and expertise.

Testing How They Participate

As we shift from crowdsourcing *widely* to crowdsourcing *smartly*, however, it is important to remember that spotting the citizen-experts capable of helping to tackle a problem is half the equation. No matter how well targeted, a group of people can produce impact only when they come together productively.

Hence we also need experiments to identify the best "horses for the courses"—not only which types of targeting, but also which types of online collaboration are most effective at different points in decisionmaking. It will be important to model and test the use of different kinds of collaborative platforms for obtaining expertise at each stage. In parallel, there is complementary work to be done on the right group size. Targeting can, after all, zero in on teams and online committees as well as individuals. How many people must be approached to get the right-sized group can be identified through experiment. In an offline study in China, researchers gathered data over many years to study the effects of group size on the outcomes of citizen monitoring of illegal activity, such as logging in the Wolong Panda Reserve. Group size had a U-shaped relation to outcomes. "Intermediate group sizes of eight or nine households were optimal in balancing between two opposing factors: free-riding (the tendency to let others in the group do the work) and within-group enforcement."[73] These findings pointed to better strategies for effective governance.[74]

Once the right audience is identified, what platforms exist to get people to remix, combine, and develop an optimal design? Software for divvying up tasks and roles has reduced the cost of such collaboration and group formation online. Still, we will want to understand better which type of group participation is most effective, when, and in which contexts. It is important to distinguish between different types of crowdsourcing and their features to perceive opportunities for testing.

As currently used, the term "crowdsourcing" can obscure these differences. When the White House claimed to crowdsource "bold ideas" to accelerate economic growth and prosperity, it meant individuals should send in emails in response to a set of questions, without deliberation or discussion.[75] This is very different from the 2012 crowdsourced Magna Carta for Philippine Internet Freedom, which citizens created in response to the passage of an anti-cybercrime law that many feared would have a chilling effect on online speech, or

Brazilian solicitation of citizen input concerning a civil liberties bill on the Internet known as the Marco Civil between 2009 and 2014.[76] In these projects, many people worked collaboratively on the same document.

Understanding the different ways that groups function online allows us to design effective ways for groups to participate in different tasks of governance. Looking at examples of online collaboration helps us distinguish between different forms of collaboration based on what people are being asked to do and conceptualize experiments for testing what works.

Consider five broad ideal types of crowdsourcing activity. The first involves *crowdsourcing ideas*. When organizations crowdsource ideas, they are outsourcing their brainstorming and soliciting suggestions for solutions to specific problems. Open innovation and ideation platforms, used to solicit ideas from a distributed population, have become increasingly common.[77] More than just digital whiteboards, these platforms enable people to submit new ideas, search previously submitted ideas, post questions and challenges, discuss and expand on ideas, vote ideas up or down, and flag ideas that do not fit the question.

Such crowdsourcing of ideas is growing more prominent in the public sector. An online game by Community PlanIt called Youth@Work Bhutan gave Bhutanese youth the ability to deliberate on issues involving youth unemployment. Players were posed questions in the form of challenges and earned points over three week-long "missions" for developing answers to address the problem. The Ministry of Labour committed to using the proposed ideas in its policymaking.[78] Increasingly, citizens are responding to *specific* requests from officials, where input is intended to inform a decision-making process.

Imagine if, in the aftermath of the 2010 BP oil spill in the Gulf of Mexico, the 2010 Big Branch mine disaster in West Virginia, and other crises, the White House or Congress (or Facebook, for that matter) had turned to an open innovation platform to tap the best scientific minds to come up with creative solutions faster. For governments, crowdsourcing ideas present an opportunity to sift through proposed pathways to solutions without requiring any long-term commitments or the wholesale abdication of decisionmaking control.

Measuring public opinion, traditionally the domain of polling, is a second area of crowdsourcing that is building momentum. The *crowdsourcing of opinions* in the form of sentiment analysis is different from traditional polling, in that this kind of collaboration aggregates large qualities of data to infer public opinion, often without the explicit participation of individuals.[79] One report found that 84 percent of best-in-class companies improved their overall performance, customer satisfaction, risk management, and actionable insights by monitoring and analyzing social media for people's opinions.[80] Sentiment analysis brings together the fields of language processing, computational linguistics, and machine learning to enable businesses, and to a lesser extent governments, to infer subjective meaning from texts on a large scale. The private sector is increasingly drawing on sentiment analysis and opinion mining—extracting opinions from reams of text—to help companies manage a brand and investors to improve decisionmaking.

It may seem that a Twitter search of a brand name or Justin Bieber or the Gaza Strip would provide a sufficient snapshot of current public opinion on the subject. However, tools for crowdsourcing feelings have grown in sophistication to the point where the "noise" of slang, sarcasm, and local expressions can be accounted for and the public's underlying opinion predicted. A study conducted by Carnegie Mellon University in 2008 and 2009 found that using sentiment analysis to analyze more than 1 billion Twitter messages for people's attitudes on consumer confidence and presidential job approval were largely consistent with the results of expensive telephone public opinion polls conducted by Reuters, Gallup, and pollster.com.[81]

Following the UK riots in 2011, Professor Rob Procter, then director of the Manchester eResearch Centre, led a group of researchers in the analysis of tweets about the riots. The *Guardian* dataset that enabled the analysis included over 2.5 billion tweets during the time of the riots.[82] The team then created a directed, unweighted network by intersecting the friends of all users with the list to better understand the sentiments of the larger community following the riots. The network enabled the analysis of how neighborhoods, interest groups, and media outlets interacted and were influenced by different social media activity. It also helped to produce an understanding of the different types of online engagement—from active leaders to listeners—and of which voices were the most influential outside of traditional interest groups. The study also identified which topics different groups talked and cared about the most.

Going forward, the crowdsourcing of opinions by the public sector could offer more sophisticated ways to gauge the policy opinions of the public, *before* undertaking or abandoning a costly project. Imagine a mayor who, despite a limited budget, plans to renovate a local park for millions of public dollars. Before starting the process, she crowdsources the feelings and opinions of her constituency and determines that, in general, people are happy with the existing park and that their biggest desire is to replace the swings in the playground.

As budgets continue to tighten across the public and private sectors, a third area is the *crowdsourcing of funds*, or crowdfunding, which is providing a new means for collecting the money needed to support specific projects. The most famous crowdfunding platform is Kickstarter, a website that lets individuals with creative projects—from full-length movie scripts to potato salad recipes—ask an audience of people to donate their own money to support the projects they like.[83] In recent years, the Kickstarter model has spread to the public sector, where the UK-based Spacehive is giving citizens the chance to fund civic projects or, as they put it, "Fund a new park or renovate your high street as easily as buying a book online."[84] Citizens are funding public projects developed by other citizens. The civic crowdfunding model, first embodied in Spacehive, has since spread around the world, including to the United States, where Neighbor.ly is helping municipalities (not citizens) leverage the crowd to fund public projects.[85]

Crowdfunding, unlike traditional slow, multistep funding processes such as grants and bank loans, allows entities with interesting projects to connect with funders quickly and directly. Micropayment platforms like Flattr radically distribute the process even further down the long tail by enabling large numbers of people to support projects with very small donations, rather than the larger contributions possible on crowdfunding sites. This connection, while accelerating the rate of funding, also forces project developers to consider the public's appetite for an initiative (and faith that it can be delivered) from the outset. Crowdfunding is a viable option only for projects that can mobilize a community of individuals with a personal investment— literally and figuratively—in the growth of the initiative.[86]

Fourth, *crowdsourcing tasks,* or "microtasking," involves distributing small bits of work or repetitive tasks to people in a participating group using a platform such as Amazon's Mechanical Turk. The TED Open Translation Project, for example, asks members of the TED community to translate and subtitle TED talks. As of 2014, the project had produced 60,210 translations in 105 different languages, written by 17,169 different translators.[87] Such a project would be far too complex and time-consuming for a subset of TED employees to accomplish alone. DuoLingo is both teaching people a language and growing the corpus of Wikipedia by asking volunteers to translate and edit the translations of Wikipedia pages one sentence at a time.

As concerns rise about the impact of automation on labor markets, microtasks offer a form of (albeit cheap) work that cannot be completed by a computer. Despite being relatively simple, these jobs require some human intelligence. Because microtasking projects often pay volunteers, this gives some participants an incentive to game the system.[88] Quality control—either in the form of post hoc review or the better targeting of participants—is therefore essential.

And fifth, *crowdsourced data gathering* is emerging as a technique for providing a more comprehensive picture of the conditions in a community.[89] Data are gathered through mechanisms such as SMS text messaging, social media, or websites and then leveraged toward a variety of ends. In Boston, Street Bump is a crowdsourcing project that draws on real-time data provided by local drivers to improve neighborhood streets.[90] The mobile application includes a motion-detecting accelerometer, which senses bumps and simultaneously records and imports the location from the phone's GPS to Street Bump, allowing government to target repair efforts toward areas with the most need and to formulate longer-term investment strategies. In Uganda, more than a quarter of the caloric intake in the national diet comes from bananas. Over the first decade of the millennium, banana bacterial wilt (BBW) has been wiping out crops, costing hundreds of millions in economic losses and threatening public health.[91] But thanks to the United Nation's mobile SMS-based reporting tool *Ureport*, the Ugandan government has been able to engage 300,000 citizens in reporting and sharing data back about the state of BBW, giving the government insight into on the ground conditions and enabling officials to communicate back to citizens together with recommendations for how to treat crops and combat the disease.

Data gathering can also help fill communication gaps during rapidly developing events. The United Nations Institute for Training and Research and the Asian Disaster Preparedness Center relied on the crowd to help in the response to crisis conditions after extreme flooding in Bangkok in 2011. Using Asign, an application that enabled accurate geo-referencing data to be collected based on volunteer citizens' photographs, flood levels could be monitored in a way that would not have been possible through the pictures alone.[92]

In all of these initiatives, information provided by citizens helped to improve decisionmaking and response to both everyday and crisis-related events. Crowdsourcing data can clearly help in evidence-based agenda setting, but the volume of collected data can present data mining, verification, and prioritization challenges. In other words, institutions seeking to leverage data gathered from the crowd need the personnel, resources, and strategies in place to make the most of that input.

In all types of crowdsourcing and across sectors, creating adequate incentives for participation is an ongoing challenge. Financial prizes can be very effective in some contexts, but are far from the only means for promoting engagement. Competition can also be a motivating factor for people interested in solving tough problems. Increasingly, gamification is bringing competitive elements and a sense of fun to crowdsourcing projects focused on easier tasks than building a 100-mile-per-hour automobile engine.[93]

In some crowdsourcing projects, people stand to gain personally even if the projects lack a financial reward or competitive bent. A desire for public recognition—thumbs-up votes, for instance—can push people to participate to the best of their ability. Skill building is another driver of participation since crowdsourcing projects, often without charge, offer participants the ability to develop and hone new skills.[94] The need for the wider use of new expertise-based credentials is embodied in this form of motivation. Attaching meaningful public markers to valuable work completed through crowdsourcing will help connect public recognition with skill-building incentives and drive even more engagement.

Personal benefits like prizes, skill building, or public recognition motivate some people. But some participants are more concerned with building knowledge than with extrinsic personal gain. Wikipedia is one example of a collaborative knowledge-building project.[95] Although such projects present an opportunity for growing and accumulating knowledge in a specific field, they are works-in-progress that need consistent feeding and maintenance. In the same vein, a desire for community building can also act as an incentive for participation. Here, it is important for the organizers of a crowdsourcing initiative to ensure that participants' efforts directly touch a common cause, like creating the best encyclopedia in human history.

Experiments with Incentives and Motivation

The fact that off-duty doctors, nurses, and paramedics will respond to a 911 call to provide bystander CPR is not surprising. Websites like GoodSam and Pulse Point are based on the assumption that you are never more than a few hundred feet from someone motivated to help. But will they participate? "You are probably never more than half a mile from a plumber," says GoodSam co-founder Dr. Mark Wilson, "but I don't think it would work in the same way." It remains to be seen whether particular groups of people are more or less likely to respond to a targeted request to participate for a public purpose. One of the key assumptions underlying smarter governance is that more people would participate if the opportunity spoke to their abilities *and* involved doing good.

"People are drawn to participate because some psychological, social, or emotional need is being met. And when the need isn't met, they don't participate," writes *Crowdsourcing* author Jeff Howe.[96] Smarter crowdsourcing projects offer the opportunity to experiment empirically with *incentives* for engagement in policymaking, using the behavioral insights we have from other fields about why people sign up and what causes them to participate in order to understand their motivations.[97] The FDA Profiles pilot searches only scientists inside the FDA, where people must participate because of their jobs. But when people are not required to participate, will they readily say yes if the request matches their expertise? Does matching, in and of itself, increase the incentive to participate? Does matching people to socially beneficial projects create a further incentive to participate?

FDA Profiles speaks to the intrinsic motivation of employees who presumably already care about medical device review, for example. Intrinsic rewards include such things as autonomy or the degree of freedom and creativity allowed by a task; being part of a community; learning during the process of contributing; and altruism.[98] For example, "tight engagement with the community was critical to the success of knowledge sharing within the Stack Overflow community."[99]

Wikipedia participants noted that contributing in and of itself creates a sense of satisfaction.[100] When asked why they contribute to Wikipedia, respondents said they did it to educate humanity and raise awareness (48.9 percent); to feel as if they made a difference (17.78 percent); and to give back to the Wikipedia community (15.56 percent). This is not uncommon: 53 percent of developers surveyed were motivated by social motivations, such as learning, sharing knowledge, participating in new forms of cooperation, and participating in the open source community.[101]

By contrast, the motivation to say yes is sometimes impelled by extrinsic considerations, such as monetary reward or gains in reputation or recognition. For example, the Stack Overflow question-and-answer website keeps track of users' participation by awarding points, which in turn are used as a measure of reputation in the community. Much of the literature on online Q&A sites suggests that, although monetary incentives such as prizes may entice people to participate, social incentives, such as interactions with contributors, lead to their continuing participation.[102]

Here we have another opportunity to deepen our understanding of the social science of how different groups respond to diverse incentives. Do they respond better to the compelling nature of a call or to an opportunity to participate that will use their expertise and know-how? Does the framing of the challenge matter more than how participation is structured and organized? Do they behave differently when a prize is offered than they do when a free T-shirt or a thank you is the only reward? How does competition, with a leaderboard showing the most frequent or helpful panelists, affect engagement?

In addition to assessing whether the nature of the reward causes or depresses participation, these kinds of experiments also provide the opportunity to look at how people behave when incentives change. One study of the use of badges on Stack Overflow found that certain users will try to game the system—altering their behavior in order to maximize their potential to receive badges. Users' votes on questions are significantly more positive before they receive the

Electorate badge than after it. This raises the question of whether—and under what circumstances—badge systems encourage high-quality contributions.[103] In another study of human motivation on social networks, the design of the interface and whether it publicly displayed user performance correlated positively with the volume of participation.[104]

For the public sector, Experts.gov therefore presents an early chance to formulate answers to a series of important research questions, including:

- What are the extrinsic (rewards, points, badges, prizes) and intrinsic motivations (altruism, autonomy, curiosity, empathy, etc.) for participation, and how do they compare within the context of solving societal problems?
- How can a single call for participation cater to the heterogeneous motivations that may be necessary to engage the optimal variety of experts?
- Are certain types of incentive structures better than others at attracting a community of expert problem solvers?
- How can institutions ensure that participants feel as though they are "legitimate, albeit peripheral" participants in the innovation process, rather than free labor whose input can be thoughtlessly accepted or rejected?

Experiments with Institutional Readiness

Designing a citizen engagement program that yields results requires focusing on the motivation of both the institution *and* the individual participant. Even if an agency can search for, find, and match experts, frame a problem, and create meaningful incentives to participate, it has to be ready and able to use what individuals contribute. This is where there is a particular dearth of research. How expertise gets used offers another avenue for exploration, especially as there exists an opportunity to adjust how people engage consistent with statutory limitations. The effectiveness of targeting has to be measured in the context, for example, of whether a regulatory review panel meets online or offline, how often, and under what rule system.

If the expertise being sought is requested online, the learning curve for new members can slow progress as new people get up to speed. Users bring experience from other interactions to new platforms. An earlier study found that a user's outlook, moderation behavior, and replies to comments are key to continued participation. Variables affecting participation outcomes include previous experience, observation, and feedback.[105] In other words, experience with online participation, rather than subject matter expertise, might be better correlated with performance.

The way an agency uses the advice it receives has to be considered when weighing how to use expertise. New York University has been conducting a set of experiments called the Open Peer Engagement Network (OPEN) to test the university's ability to govern itself more openly and collaboratively. But deploying platforms for brainstorming and discussion with members of the university community, including expert networking tools for targeting who knows what, does not ensure that the organization has the capacity to assimilate the diverse knowledge of contributing participants. People can provide input. The jury is still out on whether and how the university—any more than the FDA—will make use of it.

When the need is to identify people to serve on regulatory approval panels, review grants, or join advisory committees, institutions will probably find it easy enough to use a searchable directory of candidates to fill these predefined roles and answer clearly articulated regulatory questions. But we have yet to confirm experimentally if this is true. When the context is extraregulatory, such as university decisionmaking, there is even less certainty about the results. It is not typical in government to introduce experiments in connection with the adoption of procedural innovations, but doing so will be necessary to create the evidence, and hence the impetus, for institutions to begin reinventing themselves.

Further research is also needed to learn whether these techniques work just as well when the goals are different. For example, organizations often need quick access to missing facts. At other times, however, the need is for diverse viewpoints. Often the need is for input from people belonging to specific groups, whether stakeholders or experts. We are only at the beginning of testing when and in which circumstances targeting expertise helps and how institutions should ask for and use the input received.

Assessment of FDA Profiles should pay close attention to the impact of providing certain experts with greater access to and influence on how we govern. In the context of device review, grant reviews, and other domains where expertise is clearly called for, using expert networking platforms will likely help to democratize closed processes that rely largely on the same people. Whether this new approach privileges a new audience, however, remains to be seen. It is still an open question how to use smarter governance tools and techniques in connection with value-based—rather than fact-based—decisions.

The Center for Devices and Radiological Health is primed and eager to inject experimentation into its rollout of Profiles, cognizant of the fate that befell Aristotle. The agency has a clear idea of the kind of expertise that is needed and its urgency. But only experimentation, including a/b testing, will reveal with any certainty how Profiles worked and whether it worked well enough to merit continued investment. The FDA Profiles Experts.gov project offers the opportunity to test empirically and in granular detail whether segmenting and targeting an audience helps to improve participation.

The role of the regulator is fraught. Regulatory agencies are frequently under siege from industry groups that bemoan the burdensome and costly requirements that impede economic growth and slow innovation. At the same time, consumer groups often complain that regulators don't do enough to protect citizens from dangerous and fraudulent products and services. Both sides challenge agencies' ability to impose requirements, questioning their authority and practices, as well as their efficiency and effectiveness. The FDA has long come under attack for failing to do its job well.

Traditionally, industry and consumer groups have pushed for legislation to clarify rules and requirements and devise new procedures. New laws lead agencies to make regulations pursuant to administrative law developed in the 1940s, which offers only limited opportunity for public consultation. We make these decisions in the same way we always have, treating the legislative battlefield as the place where disputes get played out. No data are shared, and decisionmaking gets done in order to protect corporate secrecy and safeguard the integrity of the process.

Increasingly, "experimentalism is no longer confined to formal scientific labs," writes Charles Leadbeater. "It has become an organising method for social policy, startup businesses, venture capitalists, tech companies and the creative arts. Everyone it seems wants to experiment their way into the future and to do so they want labs, which are proliferating well beyond their traditional habitat in the natural sciences."[106] There is now a burgeoning movement of so-called public labs—explicitly experimental organizations with ties to government institutions. Though diverse in their approaches, these innovation labs engage primarily in ethnographic processes of engaging citizens through observation and interviews in the design of public services.[107] Most of the public labs are not focused on new technology, nor on comparative testing. But they do bring a spirit of experimentalism to public sector institutions.

This burst of enthusiasm for experimentalism is the outgrowth of the existence of technologies for collecting, processing, visualizing and transmitting large quantities of data. This new observational data is dynamic, not static. We can study patterns as they ebb and flow, such as how people create new ties, join new groups, or share content across a network.[108] Indeed, "driven by new sources of data, ever-increasing computing power, and the interest of computer scientists, social science is becoming a computational discipline much as biology did in the late 1990s," writes sociologist Duncan Watts.[109] The new availability of vast quantities of data, such as the logs of a whole country's phone system or the output of an entire social networking platform or the records of all the taxi rides or noise levels in a city, is changing how we do social research and opening up new horizons for research on citizen behavior in the wild.

Action research experiments like Experts.gov, the work Lakhani did with Harvard Catalyst at Harvard Medical School, and the work of groups like the Tobin Project and the Poverty Action Lab are not neat or perfect. It is very hard to construct perfect sample sizes with the requisite diversity in live experiments. Creating control groups and running multiple parallel trials doesn't work when not enough participants show up. In the study of institutions, there are myriad moving parts that are shaped by variables like power and money and influence that cannot be abstracted from if the goal is to understand what works in the real world and to develop "an epistemology based on practice as the fundamental unit of analysis," where action research complements rather than replaces research in the lab.[110]

New technology has finally opened up the possibility to improve how we make decisions and govern. Managed sensibly, the Experts .gov pilot can be a radical experiment. It can change the agency's default process for making by bringing in diverse expertise at the outset. It has the potential to do more when other agencies adopt it as well. The hope is that matching the right experts to the right opportunities in the right way will lead to faster and better decisionmaking. But the jury is still out.

By conducting systematic inquiries into group and organizational phenomena, such as how institutions make decisions and solve problems, action research offers a pathway for solving social problems. This demands rigorous new methods and guiding principles at the intersection between action science, network science, and human/ computer interaction. That is why we need many more interventions such as Aristotle, Peer to Patent, and FDA Profiles to help unpack the relationship between participation, knowledge, and outcomes as well as the comparisons between them.

Instead of a virtual lab in which real-world conditions are simulated on an instrumented platform where behavior is observed and measured, we need real-world "empirical and agile" action research laboratories where we can try "crowdsourcing smartly" in multiple contexts and across multiple countries, states, and cities. Many such pockets of such large-scale experimentation in the wild already exist in the business arena, where Internet-based businesses can easily undertake randomized controlled trials by manipulating their customers' experience. Citizen research lags behind.

The research on large-scale, controlled targeting experiments in the commercial world is relevant here because the conditions are "live" and focus on how to segment and target specific populations. Consumers increasingly use mobile devices to locate information and make purchases. Revenue from such purchases is estimated to exceed $100 billion by 2017. As mobile phone access expands, reaching potential customers almost anywhere, retailers must decide how to invest their advertising dollars. For example, does it matter whether a consumer is targeted with an ad on her mobile phone or her PC or tablet? Are consumers more or less likely to respond to mobile promotions in crowded environments, such as a rush-hour subway commute? Does offering a discount affect purchasing behavior, and does it matter whether the buyer is near the store whose promotion is on offer?

Thanks to advances in technology, precise targeting—such as delivering an ad to one person at a specific distance from a certain kind of store—is possible. But so, too, are natural, large-scale, randomized field experiments. Businesses are able to gather and share enormous quantities of data about potential consumers. Acxiom is a twenty-first-century Madison Avenue advertising agency. Its mad men are computer and data scientists—not graphic artists—who compile vast databases of information from Facebook, credit reporting companies, and state agencies to enable highly targeted audience segmentation. Acxiom's goal is to help its clients reach "everyone who needs to be reached."

The ability to automate the time and location at which an ad is delivered is making it possible for a mobile service provider in China, for example, to experiment by sending a short message service (SMS) to 10,690 of its mobile users chosen at random on crowded subway trains. The more dense and crowded the spaces, as measured by the number of mobile signals, the more likely people are to buy. Professor Anindya Ghose of the Stern School of Business at New York University and his research collaborators found that an average one-unit increase in crowdedness can raise the odds of mobile purchases by 12.1 percent. In packed subway trains, for example, the purchase rate was nearly 46.9 percent higher than in uncongested ones. "In a crowd, commuters perceive a loss of personal physical space, so they cope by escaping into their personal mobile space. . . . Commuters may compensate for this threat to their feeling of ease by escaping into their personal world."[111]

Another study in Germany analyzed the impact of geographic distance between a user and a retail store, in particular the interplay between the distance of the store from the viewer and the size of a coupon discount. A fourteen-week experiment comprising 354,662 observations from 3,965 unique users found that smartphone users prefer coupons from stores nearby. Around the mean, increasing the distance by one kilometer decreases the probability of using a coupon between 0.7 and 1.5 percent.[112]

Given the attention to the question of how to match eyeballs to ads and get consumers to engage with retailers, could a fraction of that research and attention be diverted to focusing on citizens rather than consumers, and on opportunities to engage in the life of our democracy rather than on buying toothpaste? In order to transition from closed to open ways of governing, we want to be able to connect citizens to opportunities to participate at least as well as we can match consumers to advertisements. That requires testing to understand what works. What does current research on commercial targeting and audience segmentation teach about how to improve the ways government engages with citizens? In what ways might technology help match citizens to opportunities to participate? And what do these experiments in the wild suggest about the opportunity to do similar research with public institutions that could change how we solve problems and make decisions?

Although this kind of testing goes on all the time in the commercial realm, once government gets involved—even if the goal of the testing is to improve the democratic experience—there will be a push to update policies on human subjects research and to address questions about disclosure and consent. Protocols will need to define what is and is not appropriate to do with and without consent. We may have to draw impossible distinctions between research designed to observe and to manipulate individual behavior. But cumbersome application and review procedures are not the answer in a world in which we can—and should—accelerate the pace of research and testing. We will need to update our thinking about institutional research practices and sharpen the definition of what constitutes the kind of human subject research that requires notice and consent.

Closed public institutions are in need of upgrades and improvements, and that means accelerating the pace of research. It will be impossible to convince politicians and policymakers, let alone the public, that the transformation of institutions is imperative without research and experimentation to prove the positive impact of such changes on real people's lives. We need to follow Mark Moore's advice to become humble "explorers who, with others, seek to discover, define and produce public value."[113]

Why Smarter Governance May Be Illegal

> Are the American people fit to govern themselves, to rule themselves,
> to control themselves? I believe they are. . . . I believe the majority of
> the plain people of the United States will, day in and day out, make fewer
> mistakes in governing themselves than any smaller class or body of men,
> no matter what their training, will make in trying to govern them.
>
> —Theodore Roosevelt, *The Right of the People to Rule*, 1912

Our government has turned to outside experts for help at least since 1791, when Thomas Jefferson, our first patent examiner, consulted with University of Pennsylvania chemistry professor Joseph Hutchinson before issuing a patent on an alchemical process for rendering seawater potable. And in 1794, President George Washington appointed an ad hoc group of commissioners to investigate the Whiskey Rebellion.[1] Since then, however, we have layered on "good government" transparency laws over the years to protect the Swiss Guard of the incorruptible professional bureaucracy from undue influence and corruption, to introduce stakeholder consensus in order to smooth the process of implementing regulations, and to centralize and strengthen political control by the White House over the federal agencies.[2]

All of these were perfectly sensible and laudable goals, but they were never designed to enhance access to expertise or encourage outside engagement. In fact, they make the techniques needed to get good expertise into government—expert networking and targeted crowdsourcing—if not illegal then so legally ambiguous as to be impractical. A triumvirate of federal statutes in force today, which ostensibly relate to how the federal government gets information to inform its decisionmaking, including how it establishes and governs the formation and running of a thousand federal advisory committees, poses serious limitations to the ability to try, let alone practice, smarter governance. The offending trio: the Federal Advisory Committee Act of 1972 (FACA), the Paperwork Reduction Act of 1980 (PRA) and the notice-and-comment rulemaking process of the Administrative Procedure Act (APA) developed in 1946. It is not an exaggeration to say that today's information laws at the federal level in the United States, especially the Federal Advisory Committee Act and the Paperwork Reduction Act, forbid robust citizen engagement.

Notice-and-comment rulemaking under the APA limits outside comments on already formulated regulations through a highly technocratic process. It attracts substantive participation from only a small number of inside-the-Beltway professionals but allows voluminous but substance-free postcard comments, pre-printed form letters distributed by interest groups to their members. It is not designed to provide an agency with the specific expertise needed to formulate a rule in the first place. The PRA actively prevents conversations with nonprofessionals as a side effect of trying to protect the public from time-consuming requests for information by government officials. FACA, the relevant law governing upwards of 65,000 volunteer advisers, has evolved into a system for ensuring interest group representation, not knowledge promotion. In practice, FACA virtually nullifies the opportunities for smarter governance presented by new technologies.

In an earlier work I have written at great length about the well-documented limitations of notice-and-comment rulemaking.[3] The problem with FACA and the PRA, the focus here, is not just that they provide policymakers no affirmative legal basis for opening up how they work or seeking out external expertise. Nor is it that they do little to remedy the *cultural* inexperience with asking for help in the government's closed-door professional environment, where admitting "I don't know" is anathema. Nor is it just their technological anachronisms. Inserting "online" into statutes adopted before the diffusion of the commercial web or personal computer technology will not—indeed cannot—bring them up to date.

To be sure, these laws are in urgent need of modernizing: Does a government website customer satisfaction survey implicate requirements of the Paperwork Reduction Act? What does it mean to archive a government record when the record is a wiki? Does use of free and open source software violate rules against gratuitous gifts and services under the Anti-Deficiency Act? Given the ease and intimacy of online chat, do discussions of a pending rule via social media require inclusions in the official rulemaking docket under the Administrative Procedure Act?[4] Does soliciting input via Twitter contravene prohibitions against outside groups providing consensus advice to agencies?[5] Current statutory language is silent on these matters.

But the challenges these laws pose to smarter governance run far deeper than the need to update the statutory framework. Party politics and the desire by some to bend science to specific political ends are certainly obstacles. But even in the absence of bad intentions, the limitations of the current legal framework is a systemic problem rooted in a failure to perceive advisory committees as bringing expertise, as opposed to interests, to bear or to view the broader citizenry as possessing relevant intelligence.

The legislative framework that governs how administrative organizations obtain expertise through advisory committees, in particular, is designed to safeguard professional bureaucrats from abuses by outside advisers. The law is silent, however, on the epistemic practices of governance and the question of how advisory committees contribute to organizational learning. House Republicans' passage of a bill in 2014 to restrict scientists from serving on the EPA's Scientific Advisory Board in connection with matters about which they have actually done research is, of course, a reflection of a probusiness and antiscience attitude. But it is also symptomatic of the failure to view advisory committees as a vehicle for delivering expertise.[6] Today's legal structures governing outside expertise, no matter the party in power, are designed to keep outside expertise outside, not to make government smarter.

Origins of the Federal Advisory System

To prevent corruption, early rules governing the solicitation of outside expertise by government dictated that only those serving on commissions and boards created by statute could be paid—and paid modestly at that. This was intended to curtail the abuse of outside committees for the purpose of doling out favors.[7] Coming on the heels of venal, nineteenth-century practices of office-selling, to which professionalism was a response, these and other strictures on outside advisers were rooted in fears of collusion and corruption.

In the early twentieth century, business groups regularly provided advice to agencies in connection with policymaking. By some estimates, four hundred temporary industry advisory committees arose during World War I, and twice as many during World War II.

At the same time, the use of outside advisory groups proliferated in response both to increasing technological complexity and to expanding government regulation under the New Deal, as well to increased government and industry cooperation during the wars.[8] For example, the Bureau of the Budget, the precursor to the Office of Management and Budget (OMB), created the Council on Federal Reports to advise it with respect to the implementation of the Federal Reports Act of 1942. Agencies could require persons to report information to the government, but only if the requirements had been approved in advance by the Bureau of the Budget. The advisory council consisted of representatives from five leading national business organizations.[9]

In response, concerns arose about the antitrust implications of these outside advisory groups, which consisted exclusively of businessmen. The fear was that policymakers could not bring the wisest judgment to bear on policy matters unless they were "protected from pressures generated by outside experts or interested parties."[10] The Justice Department issued a ruling that, so long as industry committees only advised but did not make policy, they did not violate antitrust law. Even so, concerns about industry councils persisted, and steps were taken to ensure that decisionmaking authority would rest exclusively with government professionals and not with outsiders.[11]

Throughout this early era, industry also often created its own groups to advise agencies. A 1944 statement by the attorney general stated that "responsibility for the formation of an industry committee to advise any particular department of the government is the responsibility of that department," presaging later requirements that outside expertise must be managed, controlled, and overseen by government officials.[12] Guidelines issued by the attorney general in 1955 mandated that the agenda of a committee be formulated and provided by government. The guidelines were largely ignored.

Government efforts focused on minimizing collusive behavior that took place on these councils in contravention of a faith, rooted in the New Deal, in the independence and unassailable professionalism of public policy professionals. This progressive faith was elitist and apolitical. It supposed that agencies were in the best position to decide scientifically in the public interest. Hence policymakers needed protection against undue influence by outside parties.[13] These experts— the professionals inside government and not politicians, judges, or others—would know best what to do for society.[14] In 1951, the height of this era of technocratic professionalism within government, Harvard president James Conant also proposed that outside experts should be abjured and, instead, only professional policymakers, acting in the public interest, should serve the quasi-judicial role of "briefing" both sides on any proposal to explore the merits of competing positions. This strand of thinking stemmed from a belief that outside experts were neither trustworthy nor useful. To smart government decisionmakers, they were, for the most part, redundant, ineffectual, and superfluous.[15]

Creating hurdles to the formation of advisory committees and limiting their scope therefore made sense as a means to allow *some* consultation with elite cadres of business or the academy while safeguarding the professionalism and sanctity of the bureaucracy. One

House committee report estimated that the number of federal advisory committees had grown from five thousand to more than thirty-five thousand. There was a concern that these committees had proliferated like so many weeds, without oversight or accountability, thereby posing a risk to the independence of the bureaucracy.[16] Legislation introduced in 1957 called for uniform requirements on the use of experts. Although it did not pass, the Bureau of the Budget put out a directive, later reinforced by an executive order in 1959, to tell agencies to adopt uniform requirements for such committees: a committee's members had to be engaged in and representative of the operations of the relevant industry, serve under the active leadership of a government official, and observe some transparency with regard to meeting minutes and transcripts. President John F. Kennedy later issued an executive order that capped the lifespan of advisory committees at two years. But they continued to proliferate.[17]

Then, in the late 1960s and 1970s, with the emergence of the Great Society and the intensification of interest group politics, concern arose about the representation of consumer, environmental, and labor—not just business—interests on advisory committees. As a result, the elite agency model gave way to a conception of agencies as the neutral conveners and brokers—not experts—of processes to ensure that multistakeholder interests were represented around the government's table. The Congressional Committee on Government Operations held a series of hearings on advisory committees in 1970 to explore issues of overrepresentation by business interests at the expense of the public interest, as well as accusations of waste and inefficiency. An earlier draft bill crafted in the late 1960s provided for representation by consumer, small business, and labor, and although it did not pass, it became the forerunner of the "balance" requirement for advisory committees at the center of subsequent legislation.

It was during this era that Congress enacted FACA, in 1972, to combat the proliferation of only marginally productive advisory committees with indeterminate lifespans. A default presumption of hostility to outside expertise was deeply rooted in FACA's legislative history. A special congressional study conducted in 1970 in the lead-up to the law's enactment uncovered at least 2,600 and possibly 3,200 committees. The mere fact that these groups could not be accurately counted infuriated Congress, which fulminated over the "uncontrolled and heretofore unrecorded population explosion" of committees.[18] There was a keen and mildly aggressive perception that committees, commissions, and panels were overused—all sound and fury signifying nothing: "It seems as though most Presidential commissions are merely so many Jiminy Crickets chirping in the ear of deaf Presidents, deaf Congressmen, and perhaps a deaf public."[19] Far from viewing these groups—a small number in comparison with the total size of the federal government—as necessary and beneficial to the legitimacy and effectiveness of the bureaucracy, Congress exerted pressure to reduce their number as well as to establish standards governing their creation, administration, and operation.

Taking effect two years later, FACA established procedures explicitly designed to deter the creation of advisory groups. One such procedural hurdle mandates "fairly balanced" participation, requiring inclusion of public interest stakeholders, generally speaking, and not just business interests. If a committee *had* to be formed at all, the law emphasized balanced participation among stakeholders, not on informing decisionmaking through access to specific individuals with relevant know-how.

When the legislation was enacted, there was debate about the appropriateness of a requirement to ensure a minimum number of specific kinds of public interest members on committees that dealt with technical matters.[20] When consensus could not be reached about who should be at the table, the lawmakers opted instead for a degree of public transparency. Keep in mind that Congress had only recently (in 1966) enacted the Freedom of Information Act, opening up government records to public scrutiny. There was little to no discussion, however, about the idea that members should be "balanced" in their roles and expertise.[21] The "balance" at issue was among interests, not areas of expertise.

By trying to ensure "evenhanded" membership in groups, the statute also aimed to minimize the danger of "illicit influence on agency decisionmaking."[22] It also assumed that such even-handedness would guarantee that substantive needs got met as well. In addition to a fairness justification, the thinking was that representatives of different social groups bring unique perspectives to bear. So when a committee included a representative from industry, another from civil society, and a third from an affected stakeholder group, it had all the necessary information it needed to evaluate the likely impact of a future rule or policy. What the law ignored, of course, is that when a committee is limited to a single group of a dozen people, maximizing for even-handedness can easily come at the expense of expertise.

In 1969, Harold Wilensky wrote *In Organizational Intelligence* that advisory committees are useful for representing constituencies, testing ideas, channeling public attention, mobilizing policy support, and broadening participation in the policymaking process; but he did not mention the epistemic value of committee membership.[23] Rather, he said that advisory committees—and the new legislation to govern them—should ensure that they are broadly representative of social and economic interest groups and, at the same time, support the inviolability of professional decisionmaking. Although enabling the creation and maintenance of outside advisory panels, FACA at its origin reinforced the closed-door conception of expertise. The conception of advisory committees as bringing different stakeholder perspectives to bear, rather than bringing essential knowledge to the table, took hold.

Somewhat later, in 1990, Congress emulated and codified the multistakeholder practices of the earlier era for use in rulemaking in the Negotiated Rulemaking Act, which called for convening interest groups in a consensus-building process ahead of regulatory rulemaking to shorten timeframes and forestall conflict around enactment.[24] Under Negotiated Rulemaking, all interests affected by a potential regulation get together around the table (after having complied with FACA's rules on establishing a federal advisory committee) to hash out and agree on a regulation before drafting. The act aimed to ensure a less adversarial and more cooperative process and avoid the delay of subsequent judicial wrangling. However, within a few years, the practice had already fallen into disuse and has since hardly been used—perhaps, ironically, because convening a Negotiated Rule-making group triggers FACA's burdensome and time-consuming requirements. Even if these requirements were tolerable when setting up a multiyear advisory committee, they were much less well suited to ensuring a ninety-day rulemaking process.

Still, the primary focus of the Negotiated Rulemaking Act (or "Reg Neg," as it is known), echoes a common refrain. Reg Neg aims to build consensus among stakeholders, not necessarily to get good information into the rulemaking process. Unspoken, of course, is the assumption that, through multistakeholder consensus-building processes, adversarial interests will share conflicting and relevant information. But the metric for Reg Neg, as for FACA, centers on stakeholder representation not on expertise—on measuring the inputs, not the outputs. Consistent with this stakeholder-focused model, the flavor of judicial review also shifted during this era, as judges began to interpret FACA's language to give interest groups standing to sue over regulations on the theory that these groups represented the affected public.

The Paperwork Reduction Act

The Reagan era in many ways harkened back to the New Deal notion of technocratic control by experts, but with a new emphasis on centralizing power over agency activity by the White House instead of by agency professionals.[25] In 1980, the Paperwork Reduction Act created the Office of Information and Regulatory Affairs (OIRA) within the OMB, which sits within the Executive Office of the President, and gave it oversight over all information collected by agencies under the guise of cutting red tape and creating more efficient, smaller government.[26] Efforts to centralize and control the expanding federal bureaucracy have been a tenet of the presidency throughout the twentieth century, writes Cass Sunstein, the law professor who subsequently took over as head of OIRA: "Perhaps centralized presidential control could diminish some of the characteristic pathologies of modern regulation—myopia, interest-group pressure, draconian responses to sensationalist anecdotes, poor priority setting and simple confusion."[27]

The PRA's stated goal is to improve the quality and use of federal information to strengthen decisionmaking.[28] To that end, executive branch agencies must get clearance from OIRA before collecting information, which includes "any statement or estimate of fact or opinion, regardless of form or format, whether in numerical, graphic, or narrative form, and whether oral or maintained on paper, electronic or other media" in response to identical questions posed to ten or more people.[29] What is paradigmatically contemplated is a form like an IRS 1040 or a passport application, but the language is so open-ended that it is usually read to include any questions posed to the public.

Compliance with the PRA has been streamlined in limited ways in response to the urgent need to be able to assess customer satisfaction with and get feedback on government websites, yet it is still a complex process that can take upwards of a year to approve a form, such as a survey, especially when it relates to the substance of policymaking.[30] When the government wanted to do something as laudatory as provide free text messages with National Institutes of Health–approved medical advice to new and expectant mothers in high-poverty communities, to remind them about vaccinations or make sure they had hotline numbers in the event of postpartum depression, it needed to find an outside partner to run the program. There were many reasons to want this to be a public-private partnership, among them the PRA. To operate Text4Baby, as it was called, NIH would have needed PRA approval to ask participants their due date in order to send medically appropriate advice relevant to the developmental stage of the baby. The White House was able to launch Text4Baby quickly and gather data from participants about its usefulness because Text4Baby is run by a nonprofit and sponsored by Johnson & Johnson.

OIRA's job is to minimize the information collection burden especially on those most adversely affected.[31] To do so, the PRA requires, first, that an agency publish any proposed information collection for sixty days in the *Federal Register*. After reviewing any comments, the agency then submits its proposal to OMB for review and contemporaneously publishes a second *Federal Register* notice (on its own dime), apprising the public of how to submit comments to OMB in connection with the proposed practice. OMB undertakes its own review, with no commitment to finish within a particular time frame. The OMB director then issues every information collection a control number to be printed on the form.[32] The OMB control number is two four-digit codes separated by a hyphen. The first four digits identify the sponsoring agency and bureau, and the second four digits identify the particular collection.

The irony of implementing this long, picayune, and process-heavy statute is not only that OMB trolls the federal government looking for opportunities to cut whatever red tape blocks disadvantaged people's pathway to obtaining benefits. With only two dozen officials overseeing informational practices for the entire federal government, OIRA is too small and too lacking in staff with training in information sciences, data sciences, cognitive or design sciences, or any of the disciplines needed to systematically assess the quality of government information collections or to redesign forms to be more intuitive and appealing. The PRA, like other information law provisions of the Reagan era, serves as a mechanism of political control over agencies, which until recently, when Sunstein took steps to accelerate review of certain kinds of data collections, often had to wait years to get permission to send out a survey. Although shorter, wait times are still many months.

Not surprisingly, the PRA has instilled "fear and loathing" among government personnel; they worry that any form of citizen consultation, including the use of social media like Facebook and Twitter, will trigger the requirements of the act and thus depress the incentive to consult publicly. OMB eventually recognized this, and OIRA issued guidance to clarify that the statute did not apply to traditional notice-and-comment types of participation.[33] Still, although requests for facts or opinions as part of a general solicitation of public comments do not trigger PRA's requirements, any targeted request would implicate the statute.

Hence, if an online consultation entailed government officials posting questions to a question-and-answer site akin to Stack Exchange, there is ambiguity as to whether this would contravene the PRA since the same question is being posed to ten or more people. It is true that a question appended to a rulemaking as part of a citizen consultation is not subject to the PRA, but those questions are open-ended and general. They would not yield effective responses, especially on a question-and-answer site such as Quora or Stack Exchange. To be truly effective on a question-and-answer site, a query has to be a short, concrete, specific request for information. In fact, the Stack Overflow guidelines for posing good questions specifically direct users to not be vague, to stay on-topic, and to "avoid asking for opinions or open-ended discussion."[34] Questions need to be brief and capable of being replied to with short responses. Would such questions trigger the PRA's requirements?

If a policymaker were to pose a request for insights about a particular policy challenge in education or health care or foreign policy to an online group—large or small—would it constitute an information collection? If the same question were posed using different modalities of interaction, such as, first, a brainstorming platform, then a Q&A tool, and finally, a wiki, to see how people respond and which mode is most effective, would a separate PRA clearance be required for each? No one knows. It is obvious that these rules need to be clarified to ensure that the PRA does not hinder engagement.

More important, the PRA's legitimate goals can now be met in other ways, obviating the need for the statute in its current form. To do so, first OMB, which has an inventory and copy of every one of the almost ten thousand forms in use in the federal government, should use data science to identify the redundancies in data collection practices. Reducing the ten-plus billion hours of estimated burden from information collections and form-filling is a solvable big-data problem. The inventory of forms and their frequency of use should be posted as a dataset on data.gov for others to find and analyze (the data are currently available at reginfo.gov). OMB could then post a challenge on Challenge.gov to entice people to find all the opportunities to consolidate fields and forms and minimize information collections empirically. It is not hard to imagine a challenge where participants would be asked to reduce fields by 10 percent or more.

Second, OIRA, instead of relying on review by its two dozen staff, should invite cognitive scientists, graphic designers, and information architects to review and redesign the most frequently used forms in every agency. They could be asked to create the forms anew, combine forms that are frequently completed together, and convert forms into easy-to-use apps or simply better-designed forms.

Third, an analysis of the nature of the clearance requests will reveal the most frequently requested types of clearances and make it possible for OIRA and OMB to issue guidelines for good design practices, with which agencies should comply proactively. That analysis needs to be conducted publicly and openly, sharing the data about requests and wait times and inviting third parties to assist with the analysis. By treating the process of simplifying information collection as a design and data science problem to tackle, rather than a legal compliance issue, the original statutory goals can be accomplished while clearing the way for an overhaul of the statute.

Following the passage of the PRA, a 1993 executive order added to OMB's responsibilities the authority to review all regulations and to do a cost-benefit assessment of all rules.[35] The Government Performance and Results Act (GPRA), also enacted in 1993, required agencies to submit strategic plans to OMB, including program mission statements and goals and criteria for selecting and evaluating programs. The requirements ostensibly aim to ensure coordination between and among agencies but have effectively politicized regulatory processes by institutionalizing the mechanisms for control by the Executive Office of the President.[36]

Some view this shift toward a politicization of regulatory activity as an attempt to press for anti-evolution and anti-abortion politics under the guise of science.[37] What began during World War II as an intimate relationship between science and politics and thrived under pressure to win the cold war by an intellectually inclined Kennedy administration had given way to a conservative influence that sought to appease both big business and the religious right by dressing up politics as science. But even those on the left, ostensibly representing the public interest, embraced the centralizing approach as advantageous. If the White House controlled agency activity, public interest groups could more easily zero in their lobbying efforts on the White House. Remember, White House communications are privileged, whereas comments in response to an agency during rulemaking become part of the public record and get published under the APA. A letter to or a meeting at the White House could and did remain off the record—at least until the Obama administration started publishing the logs of who visited the White House, though the contents of conversations remain protected.

Though initially a creature of the Reagan White House, this centralize-and-control agenda has been perpetuated and maintained by all subsequent administrations out of entirely rational self-interest in the face of increasing partisanship. Since all regulatory activity now requires White House approval, agencies cannot undertake new work without political direction. The agencies must work in lock-step with the White House and are subject to tighter control over communications and agenda-setting. This helps to explain why, under President Barack Obama, it was so important to Republicans to slow the political appointments process in the Senate to a standstill. With no political appointees, there can be no regulatory activity.

The combination of New Deal progressivism and its faith in technocracy, Great Society multisectoralism/multistakeholderism and

its commitment to interest group involvement, and political centralization begun by conservatives and continued by Democrats out of a
party's political interest in White House control—all of these contributed to the view of outside advisers as, at least in part, a threat to the
incorruptibility or to the inviolability of professional civil servants.

An executive order issued by President Bill Clinton required
agencies to abolish one-third of their committees by the end of 1993.[38]
Consistent with the professionalized view that places little store in
outside expertise, administrative guidance to agencies implementing
that executive order affirmatively restricts the creation of new advisory committees by capping the number each agency may maintain.[39] In operation, the policy strives to maximize efficiency, not expertise. One contemporary research study analyzing data on the use
of scientific and technical, not policy, advisory committees (meaning
those with more technical members and fewer stakeholder group representatives) suggests that agency secretaries establish expert advisory committees when they question the ability of the permanent
bureaucratic staff to ascertain the relationship between a policy intervention and its effect in the world.[40] Thus it is unsurprising that
any type of advisory committee is perceived as a threat to institutional competence, and even to institutional morals.

This suspicion helps to make clear the odd relationship between
the transparency laws that emerged to govern outside consultation
and the processes for obtaining useful expertise. Congress did not enact
these laws to maximize informational inputs or the quality of decision-
related outputs. That was neither their purpose nor their intent. The
laws simply perpetuate the structure of professionalized administration by putting obstacles in the way of convening a group of people
with relevant expertise, whether online or off, and asking them for input.

Even so, today there are at least a thousand federal advisory committees formed under FACA, comprising more than sixty thousand
members, who "advise the President and the Executive Branch on
such issues as the disposal of high-level nuclear waste, the depletion
of atmospheric ozone, the national fight against Acquired Immune

Deficiency Syndrome (AIDS), efforts to rid the nation of illegal drugs, to improve schools, highways, and housing, and on other major programs."[41]

We are not alone in the use of such advisory committees. In Europe, any group of at least six public or private sector members that meets more than once to provide nonbinding advice qualifies for the European Commission's equivalent of FACA. The Commission manages around one thousand expert groups, which in total assemble more than thirty thousand experts, to advise the commission throughout the policy process. Notably, because of the small size of the European central bureaucracy, "the institution's personnel resources stand in stark contrast to its pronounced demand for specialist knowledge to fulfill its tasks properly. As the European Commission often has insufficient internal expertise to draft, implement and monitor increasingly complex regulation, it has to rely on external sources for expertise and technical know-how and to develop an understanding for the diversity of domestic settings."[42]

Although the European committee system is widely used, there is concern that it favors "powerful and elitist interests."[43] Experts have noted that, responding to such concerns, the European Commission did indeed strive to " 'democratise' the role of experts" and "boost confidence in the way expert advice influences policy decisions" by publishing guidance "solely related to the role of expertise and knowledge in EU policy-making."[44] This guidance articulated a "commitment to enhance openness and transparency in expert groups by increasing stake-holder participation and ensuring the need for 'epistemic diversity' and 'knowledge plurality' that extends beyond a narrow scientific and technical focus."[45] As in the U.S. model, the emphasis has been not on accessing relevant expertise but on creating pluralistic committees that include a broad variety of actors, including representatives from national, regional, and local governments, as well as civil society and industry.

Sheila Jasanoff, a leading scholar of science and policymaking who has studied advisory committees comprehensively, points out that a range of approaches to balancing social interests rather than epistemological views has been adopted in Europe. The Germans go to great lengths, for example, to ensure the representativeness of "socially relevant groups" in their expert bodies, which are viewed as "microcosms of the potentially interested and affected segments of society."[46]

Scholars in the United States, including Jasanoff, Mark Brown, Chris Mooney, and others, have written extensively about the deficits and failures of American advisory committees, especially their political imbalance and manipulation. Scholars debate whether advisory committees increase participation with a net positive effect or are a symptom of government dysfunction. The legal framework that governs them is designed—not necessarily successfully—to decrease corruption and collusion or, at the very least, to increase pluralism and ensure stakeholder representation while maximizing central control and management. As noted earlier, this is not at all about expertise. Instead, it actually creates legal and procedural stumbling blocks to providing "new ideas, new facts, and new analysis."[47]

How Government Gets Expertise Today

Agencies like the Patent Office and the Food and Drug Administration, which must adjudicate applications with a significant impact on health and safety, obviously need strategies for getting information into decisionmaking. As Chapter 5 showed, their practices are outdated and suffer from significant deficits of expertise. But even regulatory agencies, which formulate policy rather than policing products, are prevented by the statutory framework from gaining access to the right know-how quickly.

Consider education, a field particularly overdue for an infusion of new ideas, facts, and analysis. The sector is in crisis. Educational attainment is stagnant, and in some areas declining. The United States ranks seventeenth among the thirty-four OECD countries in math and fourteenth out of fifteen countries on Pearson's Index of Cognitive Skills and Educational Attainment.[48] In 2012, the U.S. average mathematics score (481) was lower than the average for all OECD countries (494). The U.S. ranks fourteenth in the world in the percentage of 25-to-34-year-olds with higher education (42 percent); the odds that a young person in the United States will attend college if his or her parents do not have an upper secondary education are just 29 percent—one of the lowest levels among OECD countries.[49] These are just a handful of the statistics pointing toward declining educational attainment and a widening achievement gap.

As a result of rising costs, changing demand, and disruptive technology, policy challenges are myriad and urgent, ranging from the need to revamp curricula to adopting technologies to reducing costs and debt.[50] Student debt loads are rising as government funding for education at the state level declines and universities pass along higher costs to students in the form of tuition hikes. In parallel, the feds have stepped in with newly available credit, pay-as-you-go, and income-based repayment options, shifting power from the local to the national level. Without a way to make education more affordable for students and their parents, especially for the most vulnerable borrowers, we risk mortgaging our country's future.[51]

In such a context, government clearly has a central role to play in increasing innovation, transparency, and access to education data, particularly as it relates to providing a more effective toolkit for students and families to make smart decisions about borrowing for college.[52] Strengthening community colleges and ensuring that they serve "the dual goal of academic and on-the-job preparedness for the next generation" is also an urgent political priority.[53] Developing and strengthening American STEM education (education in science, technology, engineering, and math),[54] preparing qualified and effective STEM educators,[55] and ensuring quality in online education are all on the national agenda as well. Many education practitioners, while recognizing the utility and promise of online learning, have raised concerns about the need for "real analysis of what works for students and what doesn't."[56]

Choosing the most promising policies to address these challenges will not be easy, given the socioeconomic realities facing most learners in America today. As Clay Shirky notes, "the biggest threat those of us working in colleges and universities face isn't video lectures or on-line tests. It's the fact that we live in institutions perfectly adapted to an environment that no longer exists."[57] The reality is that the "bulk of students today are in their mid-20s or older, enrolled at a community or commuter school, and working toward a degree they will take too long to complete. One in three won't complete, ever. Of the rest, two in three will leave in debt. The median member of this new student majority is just keeping her head above water financially. The bottom quintile is drowning."[58]

In tackling these issues, the federal government is not without capable and dedicated staffers, both in the White House and at the Departments of Education and Labor, who focus on education policy and workforce training. The Domestic Policy Council has at least two senior people who work on education and labor. On the National Economic Council, there is a member whose portfolio includes workforce development. Even counting the junior staffers who assist them, though, there are fewer than a dozen people who focus on education in the White House. Add to that number a few dozen senior policymakers setting direction in the agencies (capably assisted by hundreds who do the yeoman's work of writing regulations, grant solicitations, guidance, and other policy documents). In total, by any accounting, it is a very small number, especially considering the urgency of the crisis, the ever-increasing complexity of the issues, and the high expectations we place on our government to deliver workable policy solutions.

The White House, of course, has no shortage of access to smart people. Who would not say yes when called? And people are called all the time to share ideas, write proposals, draft speeches, and attend confidential briefings (though these meetings are usually as much about marketing to, influencing, and getting the validation of the advisers as they are about seeking their advice). "The U.S. government makes a big effort to reach out to important think tanks, often through the little noticed or understood mechanism of small, private and confidential roundtables. Indeed, for the ambitious Washington think-tanker nothing quite gets the pulse racing like the idea of attending one of these roundtables with the most important government officials."[59]

I have been privy to and have even organized many of these meetings, designed to get the "buy-in" and support of those "validators" within Washington Beltway circles who are likely to be called and interviewed by the media the day after a new policy is announced, or whose attendance at the meeting, because of their status in other circles, creates a veneer of acceptability for the policy being discussed. Getting oneself into the Rolodex of a journalist almost guarantees access to the corridors of power because political officials are always working hard to win favor in the court of public opinion.

Yet White House staffers, let alone agency officials with fewer connections, only "know whom they know." The usual suspects with the right credentials and social ties tend to get invited. These are the people who already have tenure and a TED talk. Although Diane Ravitch, Linda Darling Hammond, Richard Cuban, Eric Hanushek, and a handful of other "famous" education policy academics know how to connect with those in power, most academics, even in top schools and with important and innovative ideas to share, do not have access to busy policymakers in the nation's or the states' capitals.[60]

In 2012, the Department of Education had twenty-three federal advisory committees, ranging in size from five to thirty-four advisors per group. These committees provided advice on policy topics ranging from "student financial assistance" (ten members) to "measures of student success" (fifteen members) to "national technical" advice (five members).[61] Several committees incorporated regional representation by inviting the participation of education leaders from different parts of the country; others dealt with education among specific race and ethnic groups, including Hispanics and African Americans.

The advisors who participated on these education-focused advisory committees came from a variety of backgrounds: the Student

Financial Assistance Advisory Committee's ten members, for example, included "financial aid officers, students, college presidents and administrators, officers of guaranty agencies, and leaders of national educational associations."[62] These advisers came from Nevada, California, Pennsylvania, and Maryland.

The committee was chartered by statute to make recommendations to "maintain and increase access and persistence to higher education for low- and moderate-income students." This charter includes a dizzying array of topics, including requests to provide technical expertise on "student financial aid programs, and systems of need analysis and application forms"; contribute "knowledge and understanding of early intervention programs"; recommend how it can promote "early awareness by low- and moderate-income families of their eligibility for aid"; recommend ways to "expand and improve partnerships among stakeholders" to increase both the awareness and the amount of "need-based aid available to low- and moderate-income students"; collect data on regulations and their "impact on student financial assistance and on the cost of receiving a postsecondary education"; and make recommendations "to streamline the regulation of higher education from all sectors."[63] That is an enormously complex charter for ten people to tackle, especially at one roundtable discussion in 2008 and one meeting in 2010.[64]

Furthermore, across all Department of Education advisory committees, a total of only 317 experts took part. Even fewer committees—thirteen—existed in 2012 to advise the Department of Labor, and only two of the thirteen related to training, one for Native Americans (twenty members) and one for veterans (twenty-one members). As administrative law scholar Steven Croley comments, "advisory committees before and since [FACA] have until recently been largely 'Beltway' phenomena—established by and for agency officials inside Washington and with Washington-oriented memberships."[65]

With such a limited number of groups and their small size, the slow process of establishing them, and their extended duration, there is no way for advisory committees to provide meaningful, diverse, or innovative advice or information—let alone to ensure diversity or representation of social and economic groups consistent with the goals of the statute. There is a limit to the range of skills and know-how that such a small and relatively unchanging group can bring to bear. A small group of economic advisers, for example, can barely cover classical economic viewpoints, let alone bring in thinking from neuro-economics, complexity science, or other law and economics disciplines. Take, for example, the seventeen-member President's Council of Advisors on Science and Technology (PCAST), whose members include senior executives from Microsoft and Google as well as a biologist, a geologist, and a physicist who have worked on topics ranging from advanced manufacturing to education technology to big data. For every topic, at least several of the members possess no greater skill, experience, or know-how than the smart layperson. Even the best geologist is not necessarily the most appropriate person to opine on the nation's cybersecurity policy. Yet the demands of creating these committees and running them in real time in Washington, DC, demands that they be small and emphasize credentials.

Despite the balance requirement, which was meant to put an end to the industry-only groups that were common until the mid-twentieth century, outsize influence by business has persisted, especially in the environmental policy arena and other domains involving a high degree of reliance on the natural sciences. During the George W. Bush presidency, it was not uncommon to point out that the National Coal Council, made up almost exclusively of coal industry representatives, sat on the Department of Energy's federal advisory committee on coal policy and to complain that the department had adopted 80 percent of the Coal Council's recommendations. The seven EPA panels that evaluated proposed safe daily exposure levels to commercial chemicals in 2007 included seventeen members who were employed by, or who received research funding from, companies with a financial stake in the outcome.[66] In a published statement titled "Restoring Scientific Integrity in Policy Making," over sixty preeminent scientists, including Nobel laureates and National Medal of Science recipients, lambasted George W. Bush's administration for having "manipulated the process through which science enters into its decisions."[67] The scientists and outside experts who are asked to serve on advisory committees are sometimes lobbyists operating under another name.

This is neither a new phenomenon nor the fault of any one political party. Commentators have noted that FACA has not always achieved its good-government goals or maintained separation between politics and policymaking. Complaints have long abounded that the ability of agency officials to handpick or vet members increases the opportunity for political litmus testing. Federal advisory committees are criticized either for being internal vehicles for lobbying or for not doing anything about a problem. During the hearings leading up to the passage of FACA in the early 1970s, the committee heard testimony about political vetting of even the most professional members and about abuses of the rules on openness and transparency that rendered some committees, such as the Commission on the Standardization of Screw Threads, little more than government-sanctioned industry councils.[68] (In my personal experience with presidential-level advisory committees, members were carefully vetted for ethical concerns to ensure that they could not bring disrepute on the political administration.)

Although legislation pertaining to how the federal government gets outside expertise, especially FACA, has exploitable loopholes, the courts inconsistently interpret the laws. In practice, risk-averse public servants are loath to exploit them.[69]

Federal advisory committees are, of course, not the only way the administration of government gets expertise. FACAs exist within a complex hodgepodge of avenues for getting science, research, expertise, and advising into government. Perhaps hypocritically, the law permits government officials to hire and pay contractors for their expertise. Individual contractors and consulting firms like Booz Allen and McKinsey are not subject to the same rules, such as FACA, that govern expert or advisory groups. They are seen as de facto employees doing operational work, which is exempt, rather than advising, which is not. By virtue of their employment relationship, they are deemed to be accountable enough to the civil service under their procurement contracts.

The National Academies are another source of expertise. The National Academies in the United States, including the National Academy of Sciences, the National Academy of Engineering, the Institute of Medicine, and their collective operating arm, the National Research Council, produce 250 reports annually for the federal government. Ten thousand volunteers supported by a thousand-person staff do the work. Because they rely on highly credentialed and distinguished experts—extremely busy people—they need a large staff with expensive overhead. The National Academies, as the result of recent amendments, are now also partially subject to some of the FACA rules.

Finally, federally funded research and development centers (FFRDCs), including national laboratories such as Brookhaven, Argonne, Fermi, Lawrence Livermore, and Sandia, as well as organizations like RAND and MITRE, are chartered public-private partnerships under federal law that do research and provide advice. Formally established under federal acquisition regulations, FFRDCs are federally constituted R&D organizations that "meet special, long-term needs that cannot be met by existing government or contractor resources."[70] The FFRDC, a hybrid creature that emerged after World War II at the height of the cold war, was created to meet the federal need for defense experts who could collaborate more closely with government, especially in the national security arena. Following a peak during the cold war, when places like RAND were lavishly funded campuses supported by the need to address perceived combat threats, today there are 39 FFRDCs with annual federal obligations of approximately $9.5 billion (FY 2009).[71] Most FFRDCs are operated, managed, and administered by a university, a consortium of universities, or a nonprofit and are under contract to a specific federal agency to conduct research. They and their full-time employees are treated, for all intents and purposes, as federal government employees and are not subject to the usual secrecy and nondisclosure requirements or to the strictures of the Federal Advisory Committee Act.

FFRDCs, however, are largely an outgrowth of the Defense Department. There is a wide disparity between the innovation capacity of agencies like Defense, which have access to FFRDCs such as the National Labs and the wherewithal to contract with top university and corporate talent to tackle a problem, and civilian agencies such as Education and Labor, that have little to no ability to spend money on innovation research in the private sector. HUD is not using R&D to dramatically increase the affordability of housing (in contrast to Boston's housing agency, which did launch an innovation lab).[72] Nor, given the laws discussed in this chapter, do these agencies have the appetite or legal ability to expand their use of outside expertise. For them, expert networking has the potential to become a key mechanism for getting and implementing good ideas—that is, if the legal impediments can be overcome.

7 Bringing Smarter Governance to Life

In March 1932, a small group of academics, primarily from Columbia University, including political scientist Raymond Moley, agricultural economist Rexford Tugwell, and economist Adolf Berle, began counseling candidate Franklin Roosevelt on the course of the New Deal.[1] These original "Brain Trusters" recruited other creative thinkers to join their circle. Their work must be understood in the historical context of the strong ideological and policy divisions within the political classes generally, and the Democratic Party in particular, in response to the Great Depression. There were strong differences of opinion over the origin of the crisis, the right role for government, and the most effective response. Intellectually, the central dilemma they confronted was whether to reject the principles of classical economics in the face of an unprecedented economic and political crisis.

In order to secure the nomination, FDR had to navigate these divisions successfully, appeasing all camps while setting his own course for governing. But doing so required more expertise in economic matters than he himself or his inner circle of advisers possessed. Whether to appease all sides or to shore up his economic policy bona fides so as to push through a thoroughgoing rethinking of laissez-faire tenets to enable a muscular government response to the Great Depression, Roosevelt felt an acute need for new thinkers. His adviser Sam Rosenman suggested: "Why not go to the universities of the country. You have been having some good experiences with college professors. I think they wouldn't be afraid to strike out on new paths just because the paths are new."[2] The notion of creating a Brain

Trust composed of external experts was inspired by The Inquiry, a conclave of roughly one hundred and fifty academics, whom President Woodrow Wilson had assembled to prepare for the Paris peace negotiations after World War I.

Although FDR's Brain Trust did address specific questions about tariffs, inflationary policies, deficit spending, and other such issues, its more important role was to establish a framework within which to debate, develop, solicit, and refine policy. They helped Roosevelt flesh out his rough ideas for everything from national agricultural planning and securities regulation to the Council of Economic Advisers and public works projects. They did not always agree, but the insights and energy of their discussion, debate, and dialog among themselves and with the candidate did much to shape a policy platform and, later, the New Deal itself.

Policymakers still need such outside expertise—especially on the increasing number of challenging, technical issues—quickly and daily. They also need the access to the hard-hitting conversations that the Brain Trust afforded. Today, as the gap widens between what information technology can do and the capacity of public officials to make use of it, there is a pressing need for the kind of smart, outside help—the robust dialogs between citizen-experts and government policymakers—that Roosevelt's Brain Trust represented. Now as then, good ideas that government needs are out there, including within the bureaucracy itself, among credentialed experts at universities, and more broadly distributed among those with unique kinds of knowledge.

A Twenty-First Century Brain Trust

What if, on education issues, the White House, along with agency and congressional staffers, had access at any given time to a network of leading education policy experts—a kind of knowledge network or talent bank or modern-day Brain Trust? What if this online "community of practice" were filled with smart individuals working in government and the private sector who were able to provide data, facts, opinions, advice, and insights about the pressing education challenges of the day?

Instead of having to rely on a small group of stakeholders for advice, the Department of Education could use this Brain Trust to expand their expertise, as well as their range of epistemic and social perspectives. By engaging a broad array of trans-disciplinary stakeholders working in a variety of ways together online, it could also speed up the traditional advisory committee model, where typical output now is a report or set of recommendations that takes many months or even years to produce.

This twenty-first-century Brain Trust could start out as a directory of everyone whose work touches on education policy in the *public* sector, whether in the executive or legislative branch, at the national or state level, in the United States, or in other countries' governments. Because there are so many potential data sources about civil servants and academics (including research networking tools already in use in some universities), it makes sense to begin by zeroing in, first, on their expertise. It is important to keep in mind that when we speak of citizen engagement, civil servants are also citizens. They possess relevant expertise that is, as we have seen with Aristotle and Experts .gov, not always easily accessible or findable. Simply networking these public sector professionals could dramatically expand the pool of available talent and experience across institutional silos.

As a result of experiments like the FDA's use of Profiles, the ExpertNet mandate in America's Open Government Action Plan, and pilot projects at agencies such as the U.S. Department of Agriculture, the Environmental Protection Agency, the Smithsonian, and the National Institutes of Health, efforts at building expert networks for public sector professionals have begun.

The World Bank started keeping track of the skills of its quasi-public servants via a web-based platform called SkillFinder in 2014. SkillFinder is intended to help World Bank staff, distributed around the world, to find expertise within its ranks. By combining highly structured institutional records such as human resources data about credentials with free-form user narratives and tagging and third-party endorsements, SkillFinder tracks specialization, work and project experience, languages spoken, publications, and other expertise for its sixteen thousand employees, consultants, and staff. Although originally intended to help staff identify internal peer reviewers, in its first year people are primarily using the tool to search those they know within their own departments, identifying skills gaps and strengths in order to inform plans for reorganization.

The Governance Lab at New York University has started the Network Innovators (NoI). NoI is a mobile application collaboratively designed and developed by the GovLab in partnership with governance innovation leaders across seven countries, including Mexico, Chile, South Australia, and the United Kingdom. The tool makes the know-how of government innovators on such topics as open data, prize-backed challenges, and crowdsourcing within and across governments searchable. By answering questions about one's governance innovation skills and experiences, the app creates a profile for the user, matching her to those with complementary knowledge—either those who are similar or different—to enable mutual support and learning. NoI looks beyond traditional credentials to capture indicia of real-world skills. However, instead of rigid categories or open-ended tags, Network of Innovators attempts to get at what people know by asking them the kinds of questions they could and would like to answer. In response to a series of questions, they can specify what kind of expertise they have and want to share: whether they have the ability to do something, to tell someone about it, or to refer them to others knowledgeable about topic.

From Network of Innovators, it is a small step to envision expert networking on education policy topics between the Departments of Labor and Education and among those concerned with STEM education within Energy, Defense, and the Smithsonian. Domestic innovators, of course, would benefit from access to their foreign counterparts, such as Pasi Sahlberg, director general of the Centre for International Mobility at the Finnish Ministry of Education and Culture in Helsinki and Finland's unofficial education ambassador, or Chan Lai Fung, permanent secretary of the Ministry of Education in Singapore, or Andreas Schleicher who heads the OECD's PISA comparative educational assessment survey.

From Public to Private

In addition to a public sector network, imagine expanding the Brain Trust to the academic sector. The platform would catalog and make searchable professors at schools of education who conduct research and write about the field, in the United States and abroad. It would include those who focus on comparative educational systems, like David Hogan at the University of Queensland, who studies Singapore's first-ranked system; and education policy and history experts like Diane Ravitch at New York University. These experts would not all teach at Stanford, Harvard, Wisconsin, or any of the ten top schools of education in the United States.[3] They would come from around the country and around the world, bringing with them deep local and regional expertise available in no other way.

Individually, participants might not have the ear of policymakers or a clear pathway to reach them. But they would have what those policymakers most need to hear. The network would also include economists and sociologists in graduate and business schools who study the economics of higher education and the relationship between education and innovation in the economy. It would include design professors who are reimagining the future of the university, such as the architect Ann Pendleton-Julian at Georgetown or Dan O'Sullivan, a physical computing expert at New York University. It would include computer scientists like Luis von Ahn, Carnegie Mellon professor and creator of DuoLingo; and John Mitchell, vice provost at Stanford; as well instructional technologists, such as the designers of Khan Academy or the team at Clever.com.

Done right, the provosts of a handful of founding universities would come together as a public service to start and fund an academic policy Brain Trust. The network they build would not be limited to academics and public sector professionals. They would cast their net wide to catch industry practitioners and clinical faculty with hands-on experience with what works and what does not in distance education. It would catch community college administrators and faculty with deep experience and commitment to teaching at nonelite institutions—and an even deeper understanding of the pressures faced by students who need to work and raise a family. Such people are rarely, if ever, invited to serve on Beltway advisory committees alongside CEOs and Nobel prize winners. Nor do graduate students, whose dissertation research requires them to become genuinely expert with the contemporary literatures of narrow fields of practice and who are, therefore, often more up to date on the latest research than their teachers.

A highly regarded independent group such as the National Research Council, parent of the National Academies, could be a founding partner, but its enormous professional staff and physical presence in Washington means its participation would come at an unaffordable cost. As academics and policymakers often like to quip: "We're expensive, but at least we're slow." To be effective, the Brain Trust needs to be agile and networked and open to all who are willing to participate actively. To be legitimate, it must be open to those with formal credentials and status, like the National Academies, but also to those with relevant skills and experience. People who have taught in or run high schools or who have developed online learning websites and software, as evidenced by their GitHub profiles, or who are active in the Maker Movement for hands-on learning or have entrepreneurial savvy *and* a commitment to education policy—such people would need to be invited and included. So would engaged parents and students. So would experts in education finance. And so would knowledgeable reporters and commentators.

Building a Directory of Directories

Populating a network with organizations and groups comprising either public sector personnel who work on education-related topics or university professionals with skills and know-how on education policy requires first designing taxonomies for describing relevant and innovative skills, experiences, credentials, and interests. Doing so has the added benefit of articulating the skillset needed for creative jobs in education policy.

With the right schema for organizing data about what people know and making it searchable for those who are seeking their expertise and that of their organizations, the Brain Trust would make it possible to expand the network further to include those in related businesses and industries, including technology, games, and commercial education. Anyone willing to serve the public interest by contributing actively, completing a questionnaire (with fields inspired by other expert networks like VIVO or Profiles) about his or her expertise, and signing a terms-of-service agreement requiring adherence to standards of conduct would be welcome to join.

The questionnaire would go beyond asking individual users for their credentials. Members would fill out a profile, describing in detail their relevant skills, experiences, and interests, which could include, for example, badges earned from online courses, courses taught, conferences attended, rankings on Mendeley, bibliometric scores on Academia.edu or SSRN, and other indicators of expertise. Being able to self-identify beyond publications and pedigree would help detangle and democratize participation and empower those who care about the future of education with a means of building up their confidence to contribute. Like the Network of Innovators, this Brain Trust might also ask people to describe their expertise in terms of the questions they could answer—matching the supply of expertise to the demand for it.

The actual build-out of a profile would not have to rely exclusively on user contributions. Rather, people's profiles could be augmented using existing sources of online data (and people could correct their own profiles as necessary). The Brain Trust would acquire data through a combination of self-reporting (a user filling out a profile), referral (a user suggesting someone else), and feeds (datasets from other directories or social sites) such as publications, citations, grants, and information from public and proprietary third-party databases. In education, the relevant third-party sources might also include data that shed light on time spent in a public policy position, time spent in a classroom, time spent working hands-on with young people, courses developed, and other relevant skills metrics.

Drawing on the insights of those who have studied recommender systems and "pyramiding," the network or directory would also rely on third-party recommendations of who-knows-what about various topics. In Roosevelt's Brain Trust, it was the original small group that found and included new members, such as Iowa agricultural economist and eventual secretary of agriculture Henry Wallace and Harvard Law School professor and eventual Supreme Court justice Felix Frankfurter. The more peripheral members who hailed from the academy were, in almost all cases, brought into the circle through the networking efforts of the original Brain Trusters.

Today, we have more modern means of recommending people for inclusion. Third-party endorsements and recommendations can be automated, as they are on websites like Health Tap, where doctors endorse one another for their expertise. Twitter could be called upon to help identify those who talk a lot or are perceived by others to know a lot about education policy based on discourse analysis of their social media activity. LinkedIn might share its know-how and data to ease the identification of likely candidates for inclusion.

The system would also need to optimize for the probability that the person identified will contribute productively. Hence, recommendations and personal connections (we went to school together, we were born in the same city, etc.), together with rating and ranking tools, will help create incentives for engagement and show who participates well, not just who is the most skilled. In the 1930s Brain Trust, participants were chosen not only for their intellectual prowess and gravitas but also because they had previously worked in public service and appreciated how to contribute practically. They all excelled, not just at conducting and communicating their own research, but also at sourcing, analyzing, summarizing, and communicating the research of their academic peers. These abilities made them especially useful conduits between the academic and the political domains.

Importantly, the system would not simply be a stand-alone directory—one more neglected database among many. It would be a map—a directory of directories—that enables federated searching of experts across other, existing organizations and networks of experts, such as those of interest groups like the National Education Association or the Association of University Presidents. Through agreements with other networks, opportunities to participate would be distributed broadly and widely. Members would not have to sign up at a separate Brain Trust website to receive invitations to participate. Rather, the goal would be to enable decisionmakers to push out requests, questions, challenges, and invitations to people across an array of networks. The idea is that limiting the service to public problems and public questions would motivate those who run proprietary expertise networks to make their platforms and their members available.

Even with all of the automated big-data approaches, building this Brain Trust would still depend heavily on user contributions. Collecting rich personal information from individual participants as a condition for participation will make it easier to know whom to target with specific questions. For example, a policymaker who wants insight into technology-enhanced education would want to be sure to reach people with policy experience, classroom teaching experience, and experience designing and building teaching software. In some cases, the need will be for data and facts; in others, sorting through facts supplied by others. Knowing only that someone is a member of a particular organization is less useful than knowing, as well, something about that person's lived experience, specific skills, and location. Having the ability to query the database based on specific criteria (and to test different kinds of queries) would be, in essence, a strategy for using the technologies of expertise to implement new arrangements for robust participation and advising.

Matching Supply with Demand

Matchmaking is, of course, the core of any expert networking platform. A decisionmaker seeking people to consult about a difficult question of policy would run a domain-specific query or browse the Brain Trust using filters to narrow the search. Alternatively, he could navigate the network using a visual interface such as a map of topics or geography to pinpoint the relevant organizations or the appropriate public sector project managers of public consultations in education, or legal experts in open licensing of courseware, or those with classroom teaching experience from California who are interested in education financing. Matchmaking on the Brain Trust would leverage data from the profiles with intelligent matching algorithms and visualizations.

On the Brain Trust platform, once a user has found interested peers, it would then be possible to use well-tested communications tools to organize different modes of engagement, such as brainstorming, Q&A, commenting, collaborative drafting, and both synchronous and asynchronous discussion. Communication and collaboration apps would exist "on top" of the directory so that the person convening the group could choose the type of interaction desired and even try different groups and interactions, experimenting to see what works. So, for example, a question could be pushed out via a Q&A method but also as a challenge with opportunity for more open-ended responses.

To facilitate comparative analysis, apps would be required to adhere to a set of standards regarding transparency, openness, and data sharing. The platform would be improved with rating and ranking software tools that work across apps to let people evaluate each other's contributions. In this way, the Brain Trust would build up metrics about user performance that make it easier to search for those judged to be helpful and active participants. These judgments would supplement objective, machine-generated metrics (for example, number of times someone posts, number of reposts) to ensure that evaluative metrics do not simply duplicate offline status hierarchies about credentials but, instead, rate people on the basis of their contributions to the distributed public advisory network.

Picture a government official who needs help developing a strategy for promoting hands-on science education, such as maker and hacker spaces in local communities. He needs to know: What policies can be enacted to promote the Maker Movement? What are some of the specific tactics for promoting this kind of innovation? He also needs to justify his intuition that investing time and resources on this topic matters. He wants ideas about formulating national goals related to educational achievement and workforce development, or even improvements in housing or health or crime reduction that can be realized through the creation of more maker labs in communities.

He starts by searching the Brain Trust using a set of drop-down menus and check boxes. These allow him to distinguish between those interested in the Maker Movement, those with practical experience with maker spaces, and those with relevant credentials, including badges certifying 3D printing acumen. Of course, the Brain Trust does not yet include everyone who works in advanced manufacturing or has a background in building community organizations but no perceived connection to education. But it does include education policymakers at all levels of government and in many jurisdictions. It includes university professors and students who study the Maker Movement, and it is beginning to attract active participants in the movement, such as those who run maker spaces, who have signed up either through the Brain Trust website or by invitation through the American Association for the Advancement of Science.

The policymaker would post his questions via the question-and-answer "boards" using a civic technology discussion platform such as Discourse, which enables asynchronous conversation. Discourse is set up to discuss such topics as technology-enhanced education, coding and computer science literacy, the Maker Movement, student loans and financing, guns and violence in schools, and STEM policy. In parallel, the policymaker does a search for relevant populations (all those who have checked "maker space" as part of their experience) and pushes out the questions to them to invite short responses. The question links back to the Stack Exchange Q&A micro-site on education policy where those who choose can reply. In addition, the policymaker sets up a wiki for his topic with one click to invite people to post longer case studies and stories. Again, the invitation to participate goes out to the targeted list. In this way, it becomes possible to post topics for discussion that anyone can respond to, and at the same time target participation by specific cohorts.

The platform would accommodate longer answers (such as explanations of how other cultures or countries do things). If someone wanted to empanel large or small group committees, it would accommodate that as well. Appellate lawyer and public interest advocate David Arkush, for example, has proposed that administrative agencies adopt a variation of the jury system, empaneling a thousand randomly selected citizens to offer binding advice to an agency.[4] Now imagine if, instead of convening a demographically representative sample at great expense through a polling service, you could retrieve a randomly selected thousand people from a population that has expressly indicated an interest in working on education issues. The "large jury" app would distribute an invitation to participants across the network randomly until it empanelled a thousand people.

In a variation on Arkush's idea, Administrative Conference of the United States counsel Reeve Bull, building on an idea expressed earlier by the Jefferson Center in its work on citizen juries, proposes creating citizen advisory committees—relatively small groups of citizens who would advise but not bind an agency.[5] (The Administrative Conference was created by Congress under Kennedy, subsequently defunded and dismantled in the mid-1990s, and later reestablished in 2010 by Obama to make recommendations to improve the functioning of government.) In Bull's model, participants would receive background materials generated by deliberative polling before their discussions. Bull assumes a twentieth-century and very costly model of random assignment based on telephone polling, however. Imagine that, instead of having to cajole a dozen random participants into offering insights on a policymaking process, officials could tap into parallel groups of a dozen teachers, a dozen professors, and a dozen economists, or diverse groups that mix people of varying backgrounds and expertise, and engage them in deeper conversations. Then imagine that, with the click of a button, the Brain Trust could assemble such a citizen jury automatically.

In every instance, the expert network—the directory—would make it simpler to solicit participation from a targeted pool of candidates on a nonexclusive basis. People would be selected either at random among the network or based on certain traits to ensure diversity or particular skills or experiences. Knowing who knows what in the community would also simplify the creation of different arrangements of groups within the network.

There would also be multiple ways for policymakers and Brain Trust participants to collaborate. The Brain Trust platform (which could eventually be replicated across other domains of knowledge from data science to foreign policy to telecommunications) would offer a simple way to pose and get answers to questions, perhaps using a popular Q&A platform like Stack Exchange; to convene a large advisory group informed by the Arkush model; as well as to convene small groups created at random or based on specific characteristics of sameness or difference not unlike in the Bull model. There would be a challenge platform, where anyone could set up a prize-backed challenge and advertise it through the directory, and a matching platform where people could post "want ads" to find collaborators. The platform would offer simple tools for doing all of this.

As an exclusively public sector project, the Brain Trust would reduce the likelihood of spam by restricting participation to a dotgov email address. But to be most useful, the Brain Trust would need to extend its reach into the academic sector and industry, raising the specter of excessive communications or even abuse of the platform.

To avoid overwhelming those who have volunteered for the network and possibly depressing participation, one can imagine requiring participants to earn the right to search and query the Brain Trust. Although initially, only those with a dotgov email address could convene a group and put out a request, through active participation and engagement participants would earn the privilege to ask questions of public interest and import. One could also imagine enlisting the community itself to rate and rank proposals before questions are promoted for wider distribution. Because the identity of those posting proposals would be transparent and everyone would be invited to rate the proposals, it would be harder to game the system.[6]

Experiments with the Brain Trust

Of course, experimentation will be a key part of the design of the Brain Trust. First, the platform would intentionally enable various forms of engagement, including ideation, discussion, drafting, and Q&A. It would facilitate diverse configurations of groups, including larger and smaller groups. It would enable the option to assemble specific types of people but also to opt for more or less diversity, more or less randomness, and more or less emphasis on different markers of expertise. In other words, rather than only a single algorithm for targeting, the system would enable selection according to diverse algorithms. The platform would foster testing, in particular a/b testing, by prompting organizers to select more than one mode of engagement. Hence a policymaker trying to push out a question about maker spaces to everyone who selected "Maker" as an interest in a question-and-answer format would also receive a prompt to create a randomly selected small group "jury" for a one-hour dialog.

It is important, as discussed at length in Chapter 5, to try multiple ways of soliciting expertise rather than relying on only one. At different stages of decisionmaking, different modes might be used or created to help spot issues, prioritize them, craft agendas, identify innovative solutions, develop proposals, and design experiments to test efficacy. The full network would be engaged to spot problems and smaller groups used to refine solutions. Moderation and facilitation, ongoing recruiting efforts for new members and new institutional partners, and tactics for policing abuses would be tested to see whether they improve the community. It would not matter if an organizer ran one or multiple experiments because there would be analytics running in the background to collect the results of all the experiments.

The appeal of such a Brain Trust to policymakers should be enormous. Having ready access to more diverse and enthusiastic expert-participants than is possible with the advisory committee or blue-ribbon commission system that currently exists would expand the conversation about education policy and complement the work of traditional advisory committees, associations, and unions.

Unlike the 1930s Brain Trust or The Inquiry (convened by Woodrow Wilson after World War I), whose members were hand-picked, this modern network would depend on the potential members themselves to self-select and volunteer. The questions we have to answer (hopefully informed by what we learn from Experts.gov, the Network of Innovators, SkillFinder and other projects) are: Why would they? And what incentives could be created to encourage their participation? For those participating, the Brain Trust would provide those passionate about the future of education the chance to be an online search away from being tapped by the White House, a federal agency, congressional staffers, or even the members of an advisory committee to contribute expertise in the public interest. The same reasons that motivate an academic to serve on a federal advisory committee or a public official to participate on an interagency commission would surely motivate broader audiences who have never before been asked. Prizes and other extrinsic incentives might also be employed to recruit good ideas and support challenges to further refine, develop, and implement proposals.

But there are myriad additional reasons—intrinsic and extrinsic—why a professor, venture capitalist, or graduate student would take part. Some might relish the opportunity to provide meaningful service; others would be attracted by the usual incentives of fame, fortune, fear, and fun. Academics, in particular, are motivated by fame. But there are many who will participate because of the chance to be singled out for making a useful contribution, whether by their peers, by a political leader, or by automated software-driven metrics. Harvard's Profiles expert system is in use both at Harvard and at the University of California at San Francisco. At the latter, people are eager to participate, whereas at Harvard getting people to fill out their personal profiles is like pulling teeth. The difference? At UCSF the university leadership celebrates its users and actively encourages use of the system. Celebrity goes a long way toward encouraging participation.

Some people doubtless enjoy the status that comes from being chosen for a traditional, small-group advisory committee and might wonder whether a take-all-comers platform like this is to be similarly prized. The Brain Trust would not represent a loss of the status incentive because such opportunities do not exist for most people. Data from Europe, for example, show that traditional advisory groups often include the same players repeatedly. But they represent only a tiny a fraction of the broader palette of untapped capacity in our communities.[7] The Brain Trust would reach out with much longer arms, democratizing access to expertise, and building on different, intrinsic incentives for participation. New, nontraditional advisers will appreciate the status conferred by being asked for the first time and, more important, having their feedback used and acknowledged.

Participation in the Brain Trust would be marketed to university faculty as a "perk" of employment. Participation in the Brain Trust, not unlike serving on peer review committees, would become a criterion for positive assessment in connection with tenure review. Just as on the question-and-answer community Stack Exchange, where productive participation translates into a marketable rating that a programmer can use in the Stack Exchange job marketplace, participation in the Brain Trust would become a marketplace credential that equates, if not to fortune, then to improved economic opportunities. Cash prizes will encourage others for whom the softer incentives of fame do not hold much currency.

Just as some participate in an effort to distinguish themselves, there will be others who participate for fear that not showing up will hurt them professionally. If being a Brain Truster is viewed as "the place to be" to play a role in education policymaking, some will be afraid to miss out. Especially because conversations would be federated and accessible via the websites and apps of other organizations, the Brain Trust would become a major hub for democratized participation in education policymaking.

In the end, whether people participate or not will depend heavily on design. To succeed, the Brain Trust needs to make collaboration fun, enjoyable, and lightweight. Experiments with forecasting, such as Betfair or The Good Judgment Project at the University of Pennsylvania, which test the ability of distributed networks to predict real-world events, attract participation in part because of the thrill of competition.[8] Traditional citizen engagement in policy discussions is notoriously self-serious and dull. The Brain Trust is an opportunity to combine prize-backed challenges with competitions to uncover the most relevant and useful facts and data points quickly. It can combine gamification and elements of play with more formal meetings. And it can support the desire of many to debate, argue, engage, and discover fellowship, especially among those from different backgrounds, organizations, and disciplines, who share a common commitment to improve.

A further incentive to participate would be the relevance of the conversation. Today, discussions in online venues are full of raving, counterproductive, uncivil spam. Inherent in the design of the Brain Trust is the notion that quality assurance requires transparent processes and tested moderation (and self-moderation) techniques for keeping discussions on track. Because the Brain Trust will foster collaboration on issues where the expertise is sought, it will have a high likelihood of being used in and by governing bodies. This, too, will encourage engagement. Many will be eager to do work on topics that matter to them, possibly having devoted their entire careers or courses of study to the field. Hence, even if a consortium of universities and national academies operated the platform, government must play an active role because participation needs to be relevant to actual decisionmaking.

There is no reason why only the government can use the Brain Trust. Though government must play a central role in defining the processes and procedures for its operation and establish a connection between its work and agency decisionmaking, other institutions should have access to it. Users should be able to opt in to receive communications from broader or narrower sources, and a permissioning system could give the right to post through active participation themselves.

For example, a university president or a faculty committee charged with selecting the university's approach to online education and associated issues of quality, standards, credits, remuneration, and fees could use the Brain Trust to reach relevant experts at other institutions, including tech-savvy high schools. They could also seek advice from those who have written and spoken out *against* online education. Foundation boards would use it to look for good ideas about what to fund; philanthropists could search outside the realm of the traditionally credentialed for people with innovative and exciting ideas. A leaderboard showing active and effective participation would readily serve as a useful sorting tool to identify committed people for future projects.

The Obstacle Presented by FACA

In comparison with how the government currently convenes outside experts, the Brain Trust model offers great promise for achieving smarter governance. But under our existing legal framework, if the

White House or Department of Education wanted to get the latest thinking about loan financing or community colleges or maker spaces by leveraging the kind of targeted crowdsourcing the Brain Trust approach makes possible, it would be virtually prohibited from doing so. For a Brain Trust to be relevant to decisionmaking—and hence to create the necessary incentive for participation—government must be involved. This highlights the need for strategies to circumvent or overcome the legal obstacles to expert networking and broader citizen consultation, in both advisory and operational contexts.

The stakeholder model of the 1960s and 1970s, which informs many statutes, including FACA, the Federal Advisory Committee Act, requires balanced membership. In practice, this often means the forced inclusion of groups whose points of view are already well known from their lobbying activity. It also prevents going in depth on a particular view. The President's Council of Advisors on Science and Technology, has one geologist, one computer scientist, and one geneticist—a Noah's ark of disciplines but no depth in any one of them. The statute expresses hostility to the formation of new advisory groups, largely on the basis of an earlier perception that such outside groups tarnish the unassailable professionalism of policy professionals. Even if we could convince lawmakers that FACA's good governance framework rooted in multistakeholder representation is out of date in today's world, legal impediments would still prevent establishing the Brain Trust.

The thorniest (and most often litigated) issue impeding Brain Trust–like innovations is knowing when and to which groups FACA even applies. The practical reality is that today anything that quacks like a group is treated as a statutory advisory committee by government officials and agency personnel—and thus must comply with all of FACA's good-government-focused procedural mandates, which include chartering the committee, providing advance notice, holding open meetings, and seeking a balanced membership.[9] These rules apply to any group established by statute or utilized by the president or an agency for the purpose of supplying advice and recommendations to the government.[10] FACA does not apply to groups of exclusively government officials, such as interagency working groups, but does apply to any committee that includes at least one person not employed by the government.

Consider the requirements for complying with the statute in comparison with how the Brain Trust would work. Under current law, an advisory committee can be established only after consultation with the General Services Administration (GSA) and following public notice.[11] Discretionary committees are permitted only when their work will result in the "creation or elimination of or change in regulations" or "significant improvements in services or reductions in cost" or "important additional viewpoints."[12] Likewise, if the information an agency needs is available from another source in government, the formation of an advisory committee is not allowed. The thinking seems to be that advice from outsiders is unnecessary.

Agencies can ask a single individual, a single corporation, or a single trade industry group for information or an opinion (for example, the American Bar Association for input on judicial nominations) without triggering FACA and its burdensome procedural requirements.[13] It is legal to have a standing conference call with the CEO of Google week after week. But once a group is asked for a "consensus view," FACA must be followed, making it impossible, for example, to call collectively on the heads of the major Silicon Valley firms to address the STEM crisis or a leading group of economists to discuss the student loan mess without violating the constraints of FACA.[14]

This silliness might falsely suggest that avoiding a request for consensus would skirt the statute. But the ambiguity about when FACA should and should not apply has increased restraint among agency lawyers. They fear litigation in the event consultation is deemed improper and in violation of FACA's rules on how a group is convened or how it operates. The power of the resulting constraint on innovative activity is not to be underestimated. It is not uncommon for an agency, for example, to ask an outside organization to convene a meeting or a workshop unofficially on its behalf and offsite lest the opposing political party use the occasion of the meeting to tie up the agency's lawyers in a FACA suit. FACA is thus, unsurprisingly, often criticized for its "chilling effect" on public participation.[15]

Although there is no evidence that the multistakeholder model that informs FACA leads to more effective policymaking—and plenty of anecdotal evidence to suggest the contrary—the statute makes it hard to target and convene the best experts on a topic.[16] FACA's assumption that the inclusion of all groups around a table will produce actionable advice is simply outdated.

What is clearly permitted or prohibited under FACA regarding the formation and operation of a group of experts is hard to untangle from the statutory language. It simply states that compliance with FACA's procedures is required when a committee is either "established" or "utilized" by an agency. The GSA, tasked with administering the statute, explains that where there is management or control of the group, funding of the group's activities, setting of the agenda, or any control over the group's composition by the government, FACA is triggered.[17] But this has not sufficiently resolved the law's ambiguity; FACA may still be triggered by mistakenly convening an already organized group or even by the fact that such a group works *as a group* to achieve consensus.

On one occasion, for instance, a group ran afoul of FACA by convening a group in order to hear their individual views and then, at the end, obtain a consensus view and file a joint report to an agency, which resulted ultimately in an injunction against its advice being used.[18] On another occasion, in *Nader v. Baroody,* Ralph Nader unsuccessfully challenged the White House practice of convening biweekly, three-hour meetings between senior political officials and different major business organizations.[19] The absence of evidence of an "established structure and defined purpose"—people who work in the White House often avoid committing meeting notes to paper and such notes are exempt from archiving under the Federal Records Act—spared these meetings from FACA scrutiny. But at almost the same time, a different judge, in *Food Chemical News, Inc. v. Davis,* found that two "informal" meetings (between officials of the Bureau of Alcohol, Tobacco, and Firearms (ATF) and representatives of consumer groups, and between the ATF and the distilled spirits industry) were subject to FACA. The distinction in *Food Chemical* was that the agency had called on specific persons to advise on a specific issue: "'Where there is an organized structure, a fixed membership, and a specific purpose,' courts will tend to consider such a group an established advisory committee."[20]

The Supreme Court has weighed in once about this ambiguity in the statute. "Read unqualifiedly," FACA would apply to "any group of two or more persons." The Court went on to say that Congress did not intend that result. "A nodding acquaintance with FACA's purpose . . . reveals that it cannot have been Congress' intention, for example, to require the filing of a charter, the presence of a controlling federal official, and detailed minutes any time the President seeks the views of the National Association for the Advancement of Colored People (NAACP) before nominating Commissioners to the Equal Employment Opportunity Commission or asks the leaders of an American Legion Post he is visiting for the organization's opinion on some aspect of military policy." Three justices offered a concurrence in which they emphasized that overly expansive attempts to apply FACA to groups could constitute an unconstitutional interference with the president's ability to solicit advice.[21] But this has not quashed the uncertainty about when FACA applies.

The Administrative Conference of the United States has also issued recommendations about forming federal advisory committees. As early as 1980, it commented on the ambiguity of FACA as it affects, especially, one-time group meetings. It has since called for GSA to clarify FACA's applicability: "The most serious problems regarding the coverage of FACA have involved the applicability of the Act (a) to groups convened by agencies, on an ad hoc basis, without formal organization or structure or continuing existence, to obtain views on particular matters of immediate concern to the agency, and (b) to privately established groups whose advice is 'utilized' by an agency."[22]

All this shows that, whether or not the Brain Trust would be subject to FACA, the perception that it *might* be is enough to impede any truly meaningful collaboration with government. This is enough, alas, to kill it.

Working around FACA

It is possible to imagine, however, a few simple workarounds to enable more robust yet open and transparent consultative practices like those the Brain Trust would enable. If a consortium of universities set up, funded, and maintained the Brain Trust, for example, it could be deemed to be a private association and not one "established" by government. If government officials participate but do not exclusively drive the agenda, the Brain Trust could avoid the accusation that it is "utilized" by government. In other words, it would be a private undertaking and therefore exempt.

If the Brain Trust targeted experts—but only classes or types of people, such as teachers or software developers, not specific persons by name—the Brain Trust would fall below the threshold for utilization that transforms a group into an advisory committee. The same logic would apply if people were targeted on the basis of their expertise, not their membership in any particular organization.

Because there is no prior administrative practice of automating the process of group formation and convening, there is no clearly apposite case law. There is only uncertainty, and thus hesitation. Still, there is good reason to believe that brainstorming, informal juries, question-and-answer, and other modalities of online engagement— that is, actions (1) that do not involve the physical convening of dispersed experts; (2) that do not require consensus; and (3) whose design prevents any group from expressing itself consensually—thereby avoid entanglement with the statute. Because a mere collection of individuals attending a meeting where they provide individual advice does not constitute acting "as a group" under the act, the activities of the Brain Trust could be viewed as individual advice-giving. Activities such as question-and-answer and brainstorming would escape scrutiny.

If doubt still remained, the Brain Trust could seek to register as a Federal Advisory Committee. Every small group created within it could be deemed to be a subcommittee, which does not usually require separate filing and compliance.[23] Alternatively, every time a small or large group forms online, the platform could autopublish a notice in the *Federal Register* and thereby comply with FACA's notice requirements. The *Federal Register* is the "newspaper" of the federal government, created after World War II to open a window into its workings. Otherwise, the Brain Trust groups would work openly, and their work product would be archived, complying with FACA's operational requirements.

The Federal Advisory Committee Act applies only to groups that are advisory, not operational. As a result, the National Science Foundation or the National Institutes of Health can and do regularly convene groups of peer reviewers, who work remotely as individuals reviewing a paper or grant application and get together in person to issue a consensus opinion on which papers to publish or grants to award. Because these groups do the operational work of the agency and are not advisory to other decisionmakers, they are not considered subject to FACA. Thus, if consultation were part of the definition of the day-to-day work of an agency, the Brain Trust might escape the clutches of FACA. However, a statutory pronouncement would be required to establish this as acceptable procedure.

Any one of these arguments could be expanded to make the argument academically that the Brain Trust is not governed by FACA. Still, the practical reality is that, absent a strong statement from Congress, the courts, or even GSA exempting expert networks from compliance with FACA, participation by the federal government would be limited or, more likely, nonexistent. So long as lawyers continue to understand FACA to apply to any regular or ongoing set of meetings with the same group of people, any attempt to use the Brain Trust's networked approach to ask repeated questions, convene a jury, or in any way target and organize participation, even for short duration, would be a challenge. Knowing that any meeting could accidentally be seen as an advisory committee subject to FACA rules, and more important, could trigger a lawsuit means that more than a "hack" is needed.

Why FACA Must Go

We could modernize the statute to achieve its transparency goals in today's online environment, such as by clarifying that meetings can take place online instead of in Washington. But that does not change the fact that we have no legal framework for achieving smarter governance or for developing a collaborative and consultative culture with new institutional arrangements for obtaining expertise.

If an expert or advisory group is delineated and then deemed to be subject to FACA, the group must comply with a number of statutorily prescribed procedural requirements dictating the administration and operation of the group. These requirements further limit obtaining expertise in an effort to balance participation with the good-government requirements of transparency and accountability. Although many of the operational requirements, for publication and recordkeeping, for example, are easier to fulfill with an online forum, these rules are written for an earlier, paper-based era. They did little then and do less today to foster easy acquisition of expertise.

Currently, the government is required to notify the public that it intends to form a new advisory committee and to announce its meeting dates through publication in the *Federal Register* at least fifteen calendar days before a meeting (except in exceptional circumstances).[24] Agencies pay to put statutory notices in the *Register*, which is administered by the Department of Justice, the National Archives, and the Government Printing Office.[25] If the financial disincentive were not already a problem, the Department of Justice, in particular,

234 SMART CITIZENS, SMARTER STATE

reads the Federal Register Act very conservatively to prohibit disclosures that are not affirmatively required. In fact, the Justice Department has rebuffed attempts to use the *Register* to make opportunities for the public to participate more visible (for example, publishing a calendar of online participation opportunities across federal agencies). It has done so under the theory that greater openness contravenes the purposes of the Federal Register Act, which are to give official public notice of a document, specify agency authority to issue the document, and provide the document with evidentiary status.[26]

The publication requirement under FACA to give notice of committee meetings would be problematic for the Brain Trust because it would slow down online consultations, whose appeal is that they can be undertaken quickly, especially in response to a crisis. Furthermore, absent automation, ensuring publication of all Brain Trust consultations would be burdensome, because FACA forbids simply publishing a blanket notice for all upcoming meetings or publishing far enough in advance that people have reasonable notice of their opportunity to participate. FACA's publication rules assume face-to-face meetings of the same twenty or thirty people in Washington, DC, three to four times a year, not a question-and-answer platform with thousands of participants distributed around the country.

Unfortunately, the assumptions underlying these transparency provisions are also what make the requirements so inflexible in today's age of expert networking. A group organized under the statute must articulate a purpose and a charter, which defines the scope of the committee's authority, its duties, and the estimated number and frequency of committee meetings. The charter must be published for public comment. These requirements, while favoring transparency, are inconsistent with the speed and flexibility of targeted searching to create new kinds of groups and to shift and organize people into diverse configurations on the fly.

Once constituted, a federal advisory committee has a default two-year time horizon.[27] A committee can be renewed for a second term, although renewal requires a renewed demonstration of need, lack of duplication, and a balanced membership, as well as authorization by the head of the agency that it advises.[28] Furthermore, the renewed group must file and publish a new charter before it can take any action as a group. At the same time, there are no requirements that the committee demonstrate its first-term accomplishments or failures, share any metrics, or evaluate its performance. The requirement is narrowly focused on transparency and hindering renewal, not on connecting activity or expertise to outcomes.

When it comes to operating as an advisory committee, the statute demands that a government official participate by approving or calling committee meetings, approving the agendas, attending and chairing the meetings, and reserving the power to adjourn the meetings.[29] These rules address the concern that advisory committees might "become the tail that wags the agency dog."[30] It is hard to say for certain how a court would interpret asynchronous, online activities. Beyond the challenge of determining with certainty whether FACA applies to these activities, what it would mean for a government official to run them as committee activities is unclear.

Federal advisory committee meetings must also be open to the public and offer opportunities for public participation. Congress has determined that when a federal executive official uses an advisory committee to assist him in discharging his responsibilities, in most instances he must do so openly and publicly. The group has no First Amendment right to have its communications kept secret.[31] Although the work product of adviser professionals is exempted from Freedom of Information Act requests, all other data, reports, and materials used by committees, as well as meeting minutes and all reports and recommendations produced by advisory committees, must be made public.[32] At least eight copies of each document must be filed for public inspection with the Library of Congress, with the exception of "predecisional" and "deliberative" government documents, such as draft policies or regulations provided to the committee in the course of its work.[33] Online platforms, such as those envisioned under the Brain Trust, could make compliance easier and automated.

In the offline world, FACA's openness requirements are frequently lumped together with those of the Government in the Sunshine Act, which FACA incorporated in 1977, and which requires collegial bodies such as FCC commissioners to meet in the open. The concern is that what is gained in transparency is lost in the quality of the substantive conversations. Whether the goal is to foster consensus building or gather better expertise, some contend that a degree of informality in meetings and allowing for off-the-record conversations would be helpful and would not undercut the spirit of openness at FACA's core. Philip Pettit, for instance, sharply distinguishes between the rule of reason and the will of the majority, arguing that democracy thrives when it is "not governed by public will, and often not opened to the public gaze."[34] Others have noted that "internally deliberative processes may provide an environment in which issues can be decided free from coercion and temptation and in that way approach what Habermas has called an ideal speech situation."[35]

To otherwise meet requirements of "practicing openness," federal advisory committees, like the President's Council of Advisors on Science and Technology, have experimented by webcasting their meetings and soliciting online comments, which they are free to do with no restrictions under FACA. The Administrative Conference recommends that Congress amend FACA to authorize GSA to permit online meetings, to enable "expanded electronic meeting opportunities by clarifying the extent to which 'new media' (or 'social media') can be used by agencies in connection with advisory committees," and to "illustrate, by example, options for using these media to obtain public input to agency decision-making, including development of a public consensus on specific issues that do not trigger the applicability of FACA."[36] Nothing in these recommendations, however, suggests changing the rules by which committees are created and managed. They do not argue for broadening the scope to include more people than the handful who formally serve at present.

Designing a New Legal Framework for the Contemporary Brain Trust

A new legal framework is needed—one that overcomes the anemic participation opportunities that exist for citizen experts today; that enables self-selection to the end of embracing new pathways for participation in policymaking; that encourages people to contribute their highest and best skills, experiences, and know-how to public service; and that cultivates ongoing communities of practice where citizen experts can convene and disband as needed and can engage with each other and with government. Redesigning FACA gives us an opportunity to design, test, and then build an alternative engagement model that expands our civic participation options beyond mere voting. It will make government smarter and thus better equipped to solve the tough public problems we face today.

As a practical matter, we have not yet amassed the experience and evidence needed to press Congress to change the basic framework for how government identifies and uses outside expertise, including FACA, the Paperwork Reduction Act (PRA), and the Administrative Procedure Act (APA). Without changing people's mental models of what is possible by building and operating the Brain Trust, we are not yet smart enough yet to articulate what should replace these laws.

The first step toward the necessary reform is a high-level pronouncement in support of smarter governance. Just as President Obama built on the Memorandum on Transparency and Open Government (in 2009) to issue a second memorandum (in 2013) reinforcing its commitments to open data and transparency with a new set of promises and directives, the president could now issue a follow-on memorandum reinforcing the importance of the participation prong of the original open government agenda. In that memorandum, he could elevate "smarter governance" as an administration priority and call for experimentation with new technologies of expertise to help the administration tap the intelligence and expertise of the American people and get better expertise into government to inform policymaking in key areas. The memorandum should, specifically, demand that agencies leverage the Brain Trust and networks like it in all aspects and at all times of agency operations to reinforce that FACA does not apply. Because of its ongoing, diverse, and *operational* character, the use of expert networking would not contravene FACA.

The memorandum should explain that targeting classes of people on the basis of their expertise, understood broadly, would only benefit the federal government and the public. Furthermore, it should call for a study and assessment of the quality of expertise and the impact on the outcomes of decisionmaking from expanded use of targeted crowdsourcing after two years. The memorandum should call for a prize and recognition for the agencies that make the greatest use of outside engagement.

Subsequently, to implement the new order, the Office of Management and Budget, together with the General Services Administration, should then issue guidance that calls for a two-year period of experimentation in an effort to comparatively test the workings of the Brain Trust approach. Testing should cover instructions for running comparative trials as part of consultations, including comparisons of different modalities, such as small and large juries, Q&A, and brainstorming, and further clarifying why using expert networks of government outsiders should not be deemed to contravene FACA or the PRA. This guidance should echo the call for an assessment of the outcomes that result when targeting is used to identify and tap members of specific stakeholder organizations or people with specific expertise profiles, as well as of the outcomes when targeting is based on other criteria. All data from the experiments (excluding any private information about individuals) should be made available on data.gov for anyone to study and evaluate. In addition, OMB, Congress, and a duly formed advisory committee should study the results at the end of the two-year period.

Finally, this guidance should articulate the express basis for why smarter governance is legally permissible under our existing statutory framework. Specifically, the guidance should make clear the position that targeting classes of persons on the basis of their expertise, as well as advertising to them opportunities to self-select participation in open and transparent and nonexclusive advisory platforms, does not violate FACA. The results of the two-year trial would then inform an effort to scrap existing legislation in this arena and start over with a new *smarter governance* statutory framework. We will need this legal foundation before establishing new institutional arrangements that take citizen-expertise seriously and move us toward a more collaborative and less insular way of working.

In addition to supporting the use of the educational Brain Trust with a Presidential Memorandum, Congress should charter the Brain Trust as a Federally-Funded Research and Development Center associated with the Department of Education. FFRDCs are a form of para-statal, quasi-governmental organization, akin to Fannie Mae (the Federal National Mortgage Association) and the National Park Foundation or quasi-governmental venture funds like In-Q-Tel, which invests in innovative technologies for the intelligence community. FFRDCs enjoy special, procurement-free contracting privileges with government. They can be paid with federal funds but are treated as private organizations for purposes of FACA and the like. Hence there would be no restriction on their ability to convene while enabling government employees to participate. Although FFRDCs were typically defense-related, meeting an urgent need for more expertise to support national security with ready access to scientific and technical personnel, it should be possible to make the case for why education represents a new security crisis that demands all of our attention and know-how.

The intensive collaboration with government that these changes will require cannot occur under normal circumstances. Although the classic FFRDC model assumes a physical organization with full-time employees (as exemplified by RAND and MITRE), there is no obligation or definition in the statute to force such a limitation. Rather, an FFRDC is called for where there is a special need and ability for a public-private interface to work on important and vexing problems.[37]

Hence the consortium of sponsoring universities should, in parallel, seek congressional support for the Brain Trust as an FFRDC. Alternatively or in addition, the Brain Trust might collaborate with an existing FFRDC, such as the Science and Technology Policy Institute, which works on behalf of the Office of Science and Technology Policy in the White House and, credibly, could do work on education under its current mandate.[38] The availability of an operational system under the rubric of an existing FFRDC would make it even easier for the executive branch to issue an order mandating collaboration and experimentation.

Creating the Brain Trust would eventually require legal action to push aside the impediments created by FACA, the PRA, the APA, and other statutes that, if they do not actively impede the Brain Trust, surely provide no support for recognizing citizen expertise. We need a legal framework that embraces—not only acquiesces to—citizen engagement and supports efforts to tap the intelligence and expertise of citizens. This will not be without controversy. Whether asking Congress to redo these laws or to charter a new FFRDC for the Department of Education that works in a networked way, the road must be paved by implementing the Brain Trust, seeing what works, adapting and revising it, and experimenting with new institutional arrangements for opening government to citizen expertise.

Franklin Roosevelt valued the original Brain Trust as a superior process for developing policy precisely because it was experimental and encouraged debate and dissent. His advisers naturally shared with him a certain degree of ideological commonality and commitment. But there was enough inconsistency in their viewpoints to allow for intellectual provocation, iteration, and trial and error. Through conversation and debate, the members exerted mutual influence on one another and cross-fertilized each other's innovative ideas.[39] It was an innovative gamble and the nation won.

Smarter Citizenship

> From each according to his ability, to each according to his need.
> —Karl Marx, *Critique of the Gotha Program*, 1875

Good decisions rest on good scientific knowledge. Yet experts, said Churchill, should be "on tap, but not on top." Most people recognize that it is impossible to make rational political decisions in complex societies like ours without relying extensively on expert (read: professional) advice. Despite the desperate need for high-quality expertise, its use has been the subject of much consternation and discomfort. There is a perceived conflict between expertise and democracy. Expert knowledge is portrayed as presenting a fundamental political problem for liberalism. The critique is twofold: First, the principles of equality and fairness that underpin our democracy are somehow thought to be at odds with the exclusive and exclusionary rigor demanded by the scientific and public policy professions. And second, expert administration, which keeps citizens at arm's length, is a less legitimate decisionmaking arena than the rough-and-tumble of legislatures, where values play a bigger role.

If democracy, at its most basic, is government by inclusive discussion that is largely intelligible to all participants, the professional dominance over specific categories of technological expertise confounds our understanding of legitimacy. The liberal ideal of equality demands access to the democratic conversation. The fear is that in an epistocracy—rule by the "knowers"—dominated by expert elites, including scientists with their own rituals, jargon, and customs, the experts will attempt to use their superior knowledge to exert power over others. Meanwhile, citizens who by implication lack the same knowledge will have no ability to judge what is being said or who is saying it, which means there can be no genuine discussion and no accountability.[1] The result is a distrust of "experts" as we have traditionally understood the term.

The conflict between expertise and democracy is artificial. It exists because we have long conflated "professional" with "expert" and, as a result, have separated citizenship from expertise. That mistake was possible because we were not able to make expertise manifest and searchable on a scale that mattered. Now we can. But there is baggage to contend with. Separating citizenship from expertise puts the two categories into false opposition with each other: expertise was seen to serve the values of science and was therefore, undemocratic and unaccountable. Relying on it came at the expense of listening to people and their expressions of values and opinions. As Sheila Jasanoff, a scholar of science and technology studies, has pointed out, there are two contradictory but inextricably linked arguments about why expertise threatens to delegitimize democratic institutions: expertise is too weak, and expertise is too strong.[2] There is also a third argument—namely, that the contemporary relationship between expertise and democracy is reductive and simplistic, and hence ill-suited to deal with the realities of a complex society. This chapter examines each of these arguments in turn in order to make clear the perceived legitimacy deficit that any new institutional arrangements will need to overcome. When we redefine expertise in light of new technology, we not only diffuse these tensions but also come away with a new understanding of citizenship.

Expertise Is Too Weak

In the first part of the twentieth century, the attitude toward science, like that toward professionals generally, was one of reverence for the solidity and rigor of scientific knowledge and the priest-like abilities of those who could divine the secrets of nature. Put simply, scientists were given license to pronounce, largely unchallenged.[3] During this heyday of professionalism, the view of expertise was optimistic and consistent with a belief in the orderly, rational nature of a society governed according to scientific principles. The reverence for white coats reflected a belief in a rational worldview that experimental virtuosity and theoretical brilliance could produce definitive answers to inform political opinion and policymaking.[4]

However, with broader diffusion of science and technology in daily life, the emergence of science studies and postmodern critiques of science rooted in literary deconstruction made this professional expertise suddenly more accessible and fallible. As technology became more widespread and critiques of science more commonplace, the profession of science became less remote and vaunted. As philosopher of expertise Harry Collins remarks, "On the Internet, anyone can join in the conversation about, for example, the safety of vaccines. The experience of John and Jane Doe and their children is right up there with the Nobel Prize–winning research because the Nobel Prize–winning research has been done by people like you and me."[5] The clay feet became more visible.

The view that scientific expertise was no longer infallible was reinforced by scandal and conflict, which vitiated excessive reliance on external expertise and led, with good reason, to formalistic legal restrictions intended to decrease the use of advisory committees. The twentieth century was punctuated by a series of terrible events that eroded faith in science and technology and made the potential for its abuse manifest. In the 1930s, Stalin's embrace of Lysenko and his crackpot theories of agricultural science killed off a generation of geneticists and irreparably set back the study of biology. Lysenkoism has become synonymous with the distortion of science to serve a political objective. Nazi Germany's support for the meritless science of phrenology (study of the structure of the skull) upon which to base the regime's racial policies and its criminal use of eugenics to further its campaign of "racial hygiene" and genocide was another tragic misuse of so-called science to legitimize disastrous political aims. The bombings of Hiroshima and Nagasaki, followed by the militarization policies of the cold war and the expansion of nuclear weapons science, has further brought the reduced the moral credibility of scientific expertise.

In contemporary times, President George W. Bush's notorious refusal to heed mounting evidence about man-made causes of carbon in the earth's atmosphere, as well as South African President Thabo Mbeki's rejection of the universally accepted viral theory of HIV-AIDS as a justification to slow down the delivery of AZT, especially to pregnant women, paint a picture of expertise as subject to distortion and manipulation by political masters.[6]

Whereas science during the cold war era might have seemed unfailing, today we know there are no absolute truths. "The anvil of reality has gone soft—like one of Salvador Dali's watches," remarks Collins. As a result, although it may seem obvious that policy should be informed by scientific understanding and should therefore be evidence-based, there is surprisingly little consensus about what exactly constitutes good evidence, where and how such evidence should be sought, and at what stage in the policy process different forms of evidence might be appropriate.[7] When we believe that there is no such thing as the "best science" or the "right decision" or the "perfect expert," we understandably are left with a legacy of distrust that over-reliance on fallible and imperfect expertise, especially when peddled by those vulnerable to political manipulation, is dangerous for democracy.

Expertise Is Too Strong

But there is another side to this coin: a fear that expertise will overwhelm democracy because it is so strong that it will perpetuate injustice. Behind their closed doors, experts perpetrate something *against* citizens. They form the core of an unaccountable power elite that inevitably tends toward antidemocratic behavior. Expertise compromises democratic legitimacy without a concomitant assurance of efficacy.

Take, for example, the Dodd-Frank Wall Street Reform and Consumer Protection legislation enacted in 2010. To address the specter of another "too-big-to-fail" financial firm collapse like that of Lehman Brothers, the legislation created an elite Financial Stability Oversight Council comprising the heads of the major financial regulatory agencies. "Financial markets are complex creatures upon which the rest of the economy depends; their management and optimization therefore cannot be left to the whims of the lay public, the direct commands of representatives in Congress, nor to the vagaries of market forces," comments Brooklyn law school professor Sabeel Rahman, describing the rationale behind the creation of the council. This strikes a nerve. "The market economy is at the heart of many of the most central moral concerns we face as a society: concerns about distribution, welfare, opportunity, and the good life. It is therefore also a central concern for us as citizens in a democratic polity."[8] The notion that only experts, working in insular circumstances, with other elites, are qualified to run the economy without engaging citizens offends both truth and democracy, especially when we consider that the task force has thus far failed to develop a formula to identify at-risk firms.

This complaint about the "strength" of expertise is partly a complaint about manipulation and abuse. After all, science- and technology-based industries, including pharmaceuticals, insurance, electric utilities, computers, and software, are top-spending lobbyists in the United States.[9] Obviously, scientists do not live in an ivory tower but are part of the broader economy, and that makes them susceptible to manipulation for economic as well as political gain. Giving them the keys to the kingdom can, therefore, easily result in antidemocratic abuses. No wonder we developed a legal framework to protect those who make policy from the "pressures generated by outside experts or interested parties."[10]

But the concern goes further. Yes, we want and need expertise, but the very idea of technocracy or epistocracy breeds discomfort. The fear is not that epistocracy will lead to bad decisions because the science is bad, but that it challenges the very notion of equality. If some have significantly better epistemic capabilities—relevant knowledge and skills—than others, this creates a tension with democratic equality.

"If the probability of getting 'the best or right outcome' increases when deliberation is restricted to the most knowledgeable, why not let the knowers or experts rule?"[11]

An excessive use of these professionals also places the emphasis on policy-related outcomes, which in turn lead to power further accruing to professional elites and not to citizens. Citizen engagement, in the traditional view, provides procedural legitimacy at the front end, not guaranteed better results at the back end of the policy process. Since participation does not translate directly into results, strong reliance on expertise makes output, not input, the exclusive and the excessive measure of legitimate performance. It pushes us to be more like Singapore, where there is no recognition of the inherent moral value of procedures that favor participation by all, rather than just by some. Too much reliance on experts would destroy the delicate, real-world balance in decisionmaking between democracy, which cannot dominate lest it destroy expertise, and expertise, which cannot dominate lest it destroy democracy.[12]

Expertise Is Too Modern

A third source of tension between democracy and expertise comes from those who lament that professionalized expertise has created a culture of decisionmaking ill-suited to the complexities of modern life. Imposing on society top-down rules, grounded in supposedly scientific principles, is simplistic. From the germ of Enlightenment rationality and scientific positivism springs a professional desire to define for the social world the same kind of order scientists seek in their understanding of the natural world. A belief in the ability of contemporary policymaking institutions to plan and direct society remains the core rationale of today's professionalized institutions. This belief is rooted in the experience of the industrial revolution, which rests, in turn, on a Newtonian vision of an orderly, clockwork universe driven by observable and immutable laws. If general laws direct natural phenomena in the universe, rendering life orderly and constant, there is no reason that such laws should not also apply to social institutions.

Recall from Chapter 2 how the ability to control and measure the previously unpredictable forces of time, light, and energy made possible the transition from governing institutions rooted in aristocratic privilege and personal trust to mechanistic, bureaucratic organizations rooted in professional attainment. But this transition came at a cost. James Scott, in *Seeing Like a State,* catalogs the unintended consequences: one failed attempt after another by authoritarian regimes to impose grandiose visions of social order on society. The self-confidence about scientific and technical progress, the expansion of production, and the growing satisfaction of human needs translate into a "high-modernist ideology" of social control and ordering from above commensurate with the scientific understanding of natural laws. According to Scott, however, this "imperial" scientific view cannot succeed—not because expertise is too weak or too strong but because it oversimplifies reality in an effort to control it.

British political science and complexity theorists Robert Geyer and Samir Rihani are part of a growing group of scholars who draw on an understanding of complex adaptive systems theory, which views natural systems as a balance between order and disorder and applies these insights to social systems. Like Scott, Taleb, and others, their lament is with the hyper-rationality of modern decisionmaking practices that are too mechanistic and lacking in complexity in the face of wicked contemporary problems.

This "distinctive combination of stripped-down formalism, economic calculation, optimization, analogical reasoning from experimental microcosms, and towering ambitions" characterized cold war rationality when technocracy was at its height.[13] In practice, it has clearly not helped prevent one economic or climatic catastrophe after another, thus calling into question the effectiveness—not just the morality—of the hyper-rationality at the heart of the military-industrial and military-intellectual complex.

Geyer and Rihani describe this linear, rational approach to policymaking as the adherence to four golden rules: order, predictability, reductionism, and determinism. Order is the false belief that causes always lead to effects. This is closely related to predictability and the notion that, if we know the input, we can guess the results. Both are characteristic of a kind of reductionism, which is typical for professional governance, namely that, with the right kind of training and mindset, professionals—and only professionals—can understand the behavior of the social systems they are in charge of managing. Determinism is the belief that there are clear beginnings and rational ends and the path between them can be predicted.[14]

In other words, to the extent that scientific expertise, in particular, perpetuates a rational ideology in policymaking, it is to be feared. Long before the advent of complexity theory, political theorists complained about the problems stemming from a Newtonian, rational worldview applied to the political and policy sciences. In an attempt to duplicate the search by science for universal truths supported by empirical evidence, contemporary liberal theorists, such as John Rawls and Robert Nozick, searched for all-encompassing explanatory theories about the nature of society. But as participatory democrats like Benjamin Barber and Carole Pateman make clear, these universalistic approaches assume that atomistic individuals always pursue their own rational self-interests. This is too simple—too thin a view of democracy: "The quest for certainty in political thinking seems more likely to breed orthodoxy than to nurture truth and in practice tends to promote the domination of method over substance. . . . In an attempt to mimic the hard sciences, of which they rarely have a true understanding, these social scientists have tried to subordinate every understanding of reality to some orthodox construction of understanding."[15]

These comments are representative of a wider set of critiques of the mechanistic policymaking practices that Geyer and Rihani describe, or perhaps more accurately, parody as efforts to get a poorly performing developing country, inner-city school, or diabetic patient from A to B as quickly as possible and then to C. Sarcasm aside, cold war rationality, building on New Deal conceptions of expert authority, did largely assume that rigid rules could be used to discover optimal solutions to virtually all hard problems by breaking them down into simple, sequential steps for expert analysis.[16] "The key is to do what the expert says (what others have done). . . . Emulation of previous successful strategies is the key. Experimentation, exploration and learning are best avoided and left only as a last resort."[17] The mantra was so obvious it did not need to be stated: if we get the best and brightest experts around the table, we *can* find the *right* answer. Large-scale social problems are technical problems that *can* be solved. The truth is something to be discovered by experts with the appropriate credentials.

Nonsense, say the complexity theorists. Over the long term, as painful experience shows, we cannot intervene with any predictability in the course of events. Thus, any blind reliance on credentialed, professional expertise as the key to organizing policymaking institutions is flat-out wrong. As Philip Tetlock muses in his award-winning *Expert Political Judgment,* empirical research that assesses professional political predictions against performance benchmarks shows few signs that professionals possess "greater ability to make either 'well-calibrated' or 'discriminating' forecasts" than regular citizens— or a monkey, for that matter.[18] Despite the influence and power of political pundits, accurate, long-term political forecasting beyond a year out is impossible. Pundits think they know more than they do when the clear statistical reality is that they do not. "Over the last 58 years," Tetlock comments, "there have been hundreds of studies done comparing human-based and algorithm- or machine-based prediction, and the track record doesn't look good for people."[19]

Despite hundreds of thousands of pages of "expert" academic publications, there is no assured probabilistic handle on events because the world is too complex. More research of this sort has not— indeed, cannot—add much to our capacity to act with increasing effectiveness, let alone with greater legitimacy.[20] We are doomed to make macroeconomic and other crucial decisions without any real idea of what the outcomes will be. "The remarkable absence of increasingly effective practice in these broad areas of human affairs, despite all the effort aimed at better understanding, is not a statement about the lack of *quality* in expert or scientific policy advice, but about the inherent limits on unintelligibility relevant to potential action when the future is uncertain and values conflict."[21] The world we live in today cannot be reduced to mechanistic processes capable of what Anthony Giddens terms "tight human mastery."[22] More order is not always better or more efficient or more effective.

How Smarter Governance Transforms Citizenship

Smarter governance diffuses the tension between expertise and citizenship by taking seriously the view that all citizens possess expertise. Previously, there was no way to "see" that expertise and, importantly, no way to find it. Now we can. Increasingly, rich databases of human attributes derived from datasets, individual contributions to user profiles, and third-party recommendations are making it possible to discover what people know and can do in ever more finely tuned ways. For the first time, it is realistically possible to imagine matching people in real time to opportunities to participate in governance on the basis of what they know, can do, and want to do, rather than limiting their participation merely to the act of occasional voting. For example, as Dr. Mark Wilson, neurosurgeon and co-founder of GoodSam, writes, "Using the same analogy that you are never more than 5 metres from a spider, we figured in cities you're probably never more than 200m from a doctor, nurse, paramedic or someone able to hold an airway and (if appropriate) perform high quality CPR. The problem was alerting these people to nearby emergencies."[23] Concrete examples like GoodSam, where you register that you are a trained health care worker, or PulsePoint, where you register that you know CPR, are still few and far between. But the promise is there. The technologies exist.

These technologies make it possible to go beyond the proxies of expertise like credentials or professional membership, which have led to attenuated forms of advising and a resulting distrust of experts and the governments they serve. They point to a future in which it is possible—in concrete, actionable fashion—to unlock three kinds of knowledge: expertise within government; credentialed expertise outside of government; and nontraditional forms of distributed know-how.

In the public sector, where there have been lots of examples of open-call crowdsourcing in the past few years, including among government employees at the rank-and-file level, there are now early efforts, such as the FDA Profiles project to target expertise more precisely. In 2010, to better promote science, technology, engineering, and math (STEM) education, the White House lent its support to an effort called National Lab Day, a nationwide initiative to foster hands-on STEM learning in schools by connecting volunteers to classrooms.[24] The National Lab Day website provided a place where teachers could spell out what they needed, such as help with robotics or chemistry lessons, and practicing scientists could then post profiles about themselves and what they could teach.

Notably, this grassroots effort included a call to action for *federal scientists* at the Department of Energy (DOE), National Institutes of Health (NIH), the National Aeronautics and Space Administration (NASA), the National Science Foundation (NSF), and other agencies. In a speech before the National Academy of Sciences on April 27, 2009, President Obama directed the call specifically to government employees with special expertise: "I want to persuade you to spend time in the classroom, talking and showing young people what it is that your work can mean, and what it means to you . . . to think about new and creative ways to engage young people in science and engineering . . . encourage young people to be makers of things, not just consumers of things." Many responded. Charles Bolden, NASA administrator and a veteran of four space shuttle flights, visited a local public school to build and test-fly paper rockets and to encourage participation targeted at government scientists and engineers.

More recently, in 2014, the White House started to develop "communities of practice" for mid-level innovators within the federal service and across the departments who are practitioners of citizen science or open data or crowdsourcing or prize-backed challenges. The aim was, first, to have those peers create training materials about the use and implementation of such innovations in governance. A second goal of building an expert network of people with this kind of *experience* and *skill* was to have a searchable directory available to those wishing to undertake such innovations in the future. These mid-level innovators are sometimes the hardest to find, but they are often the people with the most practical expertise and institutional familiarity with how to get things done. The International City/County Management Association, the trade industry group for city managers, operates the Knowledge Network, where local government leaders can create profiles and then post and answer each other's questions. The European Commission has created a public sector members network for experts on re-use of public sector information.

The Governance Lab is working with the National Health Service (NHS) in the UK to explore the development of a public sector data science network to help identify and make findable those within and without NHS England and the UK public service working on—or with experience in—data science in the areas of open data or big data. The goal here was to help NHS expand its open health data programs, increase their visibility and use, and better use data to track public health problems, initiatives, and outcomes. The NHS has a mandate to apply data to improve how it works, but "open data" and "innovation" teams are small and have little ability to tap into knowledgeable resources to help. Now they might be able to do just that.

The new technologies of expertise also make it possible for officials to connect people outside of government with relevant, credentialed expertise. GoodSam, which admits only credentialed medical professionals to its volunteer network of first responders, is one example. Although FDA Profiles initially aims to match government employees with opportunities to serve on medical device review panels, the intent is to expand the project to tap the intelligence and expertise of biomedical professionals in universities across the country. Chapter 6 describes in some detail the example of a proposed Brain Trust for education policy, which would connect those with specific knowledge of and interest in education policy issues to the opportunity to serve.

The County of Torfaen in southeastern Wales has been running an ongoing series of consultations in connection with the development of a community expert networking project called Wisdom Bank.[25] In light of the fact that the median age of its citizens is now in the 40s and rising, the citizens of Torfaen expressed a desire to share the wisdom and that comes from age and experience for the good of their community. There is also a high demand for skills in the region. Hence the county has been exploring how to mine the strategic asset of the experience, capacity, and desire to contribute of its population. As part of the planning process for Wisdom Bank, the community itself outlined seventeen potential uses of targeted expertise for the benefit of governing. For example, expertise could be mined to bolster employability among young people in an effort to reduce youth unemployment and reduce the demand for public services. Those with particular business acumen could staff an innovation bureau to share their experiences with building sustainable businesses and thereby promote entrepreneurship. The ideas ranged from tapping the skills of people who previously worked in large enterprises, to getting men to coach each other about health issues, to having old-timers give advice to new immigrants.

Whether it is Wisdom Bank or Pulse Point or global efforts to target volunteers, whom the French call *les herbonautes,* to help digitize the collections of the world's natural history museums, the use of targeting is on the rise to connect government with those outside its walls, especially those with craft- and experience-based wisdom. In many cases, these efforts start out as open-call crowdsourcing projects from which a cadre of "power users" emerges—those who distinguish themselves by making productive contributions—who are then targeted for ongoing participation. Experience teaches, however, that it is not enough for government to wait to see who comes in response to an open call. It must also reach out proactively to find those most likely to have the skills and information it needs.

Such targeting is an invigoration of the opportunity to participate in the life of our democracy beyond going to the ballot box. It deepens and redefines citizenship. When a person supplies information to the Patent Office in response to a pending application on Peer to Patent or responds to a targeted call to off-duty medical professionals to come to the aid of an accident victim, she is participating in governance, even if only in a small way. This has nothing to do with support for partisan causes or candidates. It has everything to do with what it means to be a citizen in a contemporary democracy.

Similarly, a focus on target-based participation in governance does not necessarily require uniform, broad-based, or egalitarian procedures. The opportunity to join a working group on educational technology via the Brain Trust might be formally open to all, but that does not mean that everyone will or should participate. The reason is simple. Most people in a given community have some knowledge about local conditions. Other kinds of knowledge, however, might belong to only a few. Targeting specific kinds of people to alert them to the opportunity to participate will, indeed, establish hierarchies of talent and ability; different people will be called upon more in certain circumstances. If experience with open source teaches us anything, it is that, in many circumstances, only a few people can and will participate well.

Assuming the ability to overcome the legal impediments, the Brain Trust would likely start out by gathering knowledge and ideas primarily from elites and, indeed, heavily credentialed elites. Although anyone could join, in reality, not everyone would. Why, then, does the Brain Trust not reinforce fears about science and democracy? Will smarter governance replace professional government with oligarchy and epistocracy?

The answer lies, first, in acknowledging that smarter governance is not rooted in the universal but anemic conception of citizenship as something associated with the act of voting and legitimated by membership in a geographic community. Rather, it aims for something quite different—namely, creating richly textured, robust, and meaningful forms of engagement between citizen, state, and society that are tailored to each person's abilities and tethered to everyday practices of governing, not politicking. What makes this kind of targeted engagement truly democratic—and citizenship in this vision more active, robust, and meaningful—is that such targeting allows us to multiply the number and frequency of ways to engage productively in a manner consistent with each person's talents, abilities, and interests.

The notion that citizens possess the expertise to participate in governing is, in fact, a classical idea. In ancient Athens, where democracy began, citizen competence and expertise were central to economic and military success. Athens, therefore, offers a useful guide for what is possible—and what works. In the fifth century BC, Cleisthenes expelled the Spartans from Athens and set in motion the building of the foundations of the democratic state. Today, modern policymaking is done largely by elites behind closed doors. Athens developed extraordinary institutional innovations for the governance— with and by citizens—of a population of a quarter million people spread across twenty-five hundred square kilometers. In doing so, it overcame the challenges of coordinating public policymaking at scale with institutional designs that made it possible to aggregate and distribute knowledge across the realm. All in all, there were fourteen unique Athenian governing institutions that managed the polis, all comprising amateur citizen participants, not professionals. The city's success, coupled with its openness and opportunity, attracted talented people from across the Mediterranean who helped populate these new institutions.[26]

Drawing on painstaking research into the inventory of assets and characteristics of one thousand Greek city-states, Stanford classicist and historian Josiah Ober explains in *Democracy and Knowledge,* the culmination of a trio of books on Athenian governance, that Athens fared better economically and militarily as a result of its participatory political institutions. Contrary to the contemporary assumption that professional organizations are more effective on a large scale, Ober shows that, in ancient Athens, these participatory ways of deciding and doing were highly effective because they were so good at making use of the city-state's human capacity to tackle its challenges. Athenian political institutions "can best be understood as a complex and effective machine designed (*ab initio* and through much subsequent experimentation) to identify and collect relevant social and technical knowledge."[27]

In ancient Athens, voting was a means to the more important end of electing one's peers and oneself to serve. Perhaps surprisingly, it was a relatively unimportant feature of Athenian democracy. Rather, the success of Athenian democracy lay in its institutions of participatory governance. Because of, not in spite of, rule by ordinary citizens, Athens was able to leverage human capital effectively and undertake complex tasks such as founding a naval base in the western Adriatic in the fourth century BC and building the sumptuous Parthenon for the equivalent of half a billion dollars.

Athens's participatory approach to public knowledge management was in large measure the cause of its economic and military success relative to other *poleis*. Athenian pluralistic institutions were able to amalgamate knowledge productively, enabling them to govern better: "Diversity of input brings to the fore latent knowledge possessed by individuals who would not be recognized by elite power holders and experts. People from various walks of life possess pertinent information and expertise that potentially promotes productive innovations," argues Ober.[28] Because Athens could take advantage of its native social capital and citizen intelligence and organize innovation-centered learning, it competed successfully over time against more hierarchical and professional rivals.[29] In Athens, he explains, engagement in decisionmaking was not a philosophical nice-to-have, but the must-have ingredient in the city-state's unique historical success.

The form of engagement practiced was the unique differentiator between Athens and other states.[30] Daron Acemoglu and James Robinson, in *Why Nations Fail,* make a similar argument while painting with a broader historical brush.[31] From the Aztecs to the Venetians to the empires of Eastern Europe, differences in economic growth can be traced to one thing, they contend. It is not culture or the plague, but institutions: institutions that are inclusive and promote the use of citizens' talent, ingenuity, ambition and ability. When rulers become extractive, seeking to oppress their people to the end of achieving economic gain, they guarantee decline. But when regimes advance human capacity, they progress.

With the sack of Athens by the Macedonians and subsequently by the Romans just before the Common Era, Athens's glorious political innovations fell victim to authoritarian domination. All later democracies root themselves in the legacy of ancient Athens (and the plebeian institutions of Republican Rome), but modern conceptions of democracy have never recovered the faith the Athenians placed in citizen self-governance—an entire system centered on volunteer teams of experts rather than on professional bureaucrats or authoritarian rulers. The inclusive institutions of ancient Athens were the outgrowth of a philosophy of citizenship that put citizens at the center and celebrated the virtues and value of *their expertise* to the success of the community. By contrast, Ober points out, "contemporary practice often treats free citizens as passive subjects by discounting the value of what they know. . . . Willful ignorance is practiced by the parties of the right and left alike." Sadly, an Athenian brought by time machine to the present would see our cloistered expert approach to problem solving and policymaking as "both worse for democracy and less likely to benefit the community."[32]

Smarter governance builds on the Athenian model that Ober describes. But the differences are as important as the similarities. The Athenians relied on techniques that left much more to chance. The Council of Five Hundred, which set the agenda for the Assembly and managed day-to-day affairs, was annually chosen by lot to serve one-year, nonconsecutive terms. With complete turnover each year, there were no old hands or insiders. True, the absence of tenure reduced corruption and prevented the creation of institutions that would attempt to exploit the people for self-enrichment. Control was diverse. Today, of course, no one would create the Brain Trust for education by dividing the U.S. population up into tribes and randomly assigning people to tribes and appointing them to serve. Random sampling, rotation, and lotteries may have worked well in ancient Athens to promote social cohesion and build national unity. But they miss what's possible today with the technologies of expertise: the pinpointing and targeting of people based on what they know, not where they live.

Smarter governance is after a somewhat different form of political economy, more akin to Marx's idea of "from each according to his ability." People need not have any expertise in politics. But if they have knowledge relevant to governing and problem solving—whether or not they are part of any credentialed elite—they will get called upon more often and given more responsibility in much the same way that meritocracies emerge in open source programming. Participation is not a free-for-all. It cannot be if it is to remain productive. Smarter governance is democratic in the same way that the production of the Ubuntu operating system is democratic. It does establish hierarchies of responsibility and talent; it does create its own forms of elites. But the opportunity is distributed, decentralized, and open to all.

Unlike voting or jury service, participation in these problem-solving communities is not restricted to citizens in the traditional sense. Anyone with a CPR certification can perform bystander CPR, thanks to Pulse Point. The Internet makes it possible to gather expertise from anyone, anywhere, whether documented or not, whether located in the jurisdiction or not. The Brain Trust, as discussed, would benefit from participation by those in Finland or Singapore, not just those in Washington, Boston, or New York. Previously, citizenship was about acts of political community like voting. Under smarter governance, citizenship entails contributing skills and information for the common good.

How Smarter Governance Transforms Institutions

Attempts, whether authoritarian or well intentioned, by decision-makers to impose a rational, technocratic, and orderly outcome on an otherwise disorderly world have arisen not just from poor judgments about what expertise is, but also from poorly designed institutional arrangements that starve public institutions of expertise and citizens of the lived experience of participating. Federal advisory committees and National Academy review panels are good at creating an authoritative set of recommendations by assembling a committee of widely recognized professionals. They review what is known about an issue, choose a likely course of action, and offer specific recommendations in keeping with the conception that there is a right answer to the complex challenge they face. But as even the executive director of the engineering division of the National Academies points out, such groups are limited in what they can do.

The National Research Council, for example, at the request of the U.S. Department of Homeland Security, convened a group of widely recognized experts in response to the failure of the electric power grid during a major blackout. They did the usual: review what is known; consider likely courses of action; and offer operational and technical recommendations to improve performance and reliability. What they did not do—because they were not well suited to do it—is "explor[e] the technical as well as societal, environmental, economic, regulatory or other broad implications of alternative scenarios of evolution of the nation's electric utility industry." Nor were they suited to explore major step-change alternatives for the future: centrally controlling the electric supply grid, for example, or fully deregulating the wholesale and retail segments of the industry.[33]

In a comprehensive 2014 study of the effectiveness of a hundred presidential and congressional investigative commissions with weighty oversight mandates, policy scholar Paul Light offers a similar set of observations. In theory, a good investigation examines "the widest range of potential causes and consequences of the crisis at hand." But of the investigations he studied, Light found that only thirty-nine were designed for inquiry into systemic issues.[34] Worse, since the investigatory trigger is often a high-profile news item such as the financial collapse in 2008 or the Solyndra scandal in 2011 (involving the failure of a much-touted solar-panel startup that received millions in federal loan guarantees), spotlights dominate, and the quality of the work can be poor and lacking even in basic fact-finding. Worse yet, these efforts often produce no results. Despite widespread attention to the BP oil spill in 2010, an investigatory commission and other advisory committee work, there has been no congressional action. The leaks continue.

The same complaint about ineffectual top-down planning gets leveled against "experts" in the field of development as well. The claim of expertise—that we know what is best for someone else—often masks a highly reductive claim to power. An expert trained in a professional discipline offers a diagnosis of a problem afflicting the poor. Trained to "draw boxes around human problems," he or she transforms them into technical issues capable of being solved. But the solutions rarely work, and when they do, often create new problems in their wake.

Tackling problems with technical interventions but absent any consideration of politics, context, and history does not address root causes.[35] Problems morph and persist because the root causes, more often than not, are political, not technical. With each reductive techno-fix, a new set of problems is created. For example, efforts to eradicate the malaria-carrying mosquito in Egypt in the twentieth century treated the problem with the insecticide DDT at considerable social and economic cost. The mosquito problem arose, however as a result of a dam on the Nile River, the economics of cash crops like sugar cane, and the spread of the railway. Each one of these was intended to develop Egypt but, in turn, led to another problem.[36]

If we believe that we live at a time of tumultuous and unpredictable change, if we believe the world is a complex system, or if we simply lack faith in top-down, one-size-fits-all planning models, then we know that major problems cannot be "solved." Our decisionmaking apparatus, however, is presently configured to offer solutions, not an evolving capacity to tackle issues as they morph and change. Reconfiguring them to target, enlist, and harness flexible coalitions of citizen expertise helps address this problem. But how can we best develop the institutional mechanisms to do this and to make good use of the results?

Orderly, mechanistic approaches to decisionmaking, as currently configured, are not designed to provide flexible, real-time responses to rapidly changing conditions. They are not designed to translate the best technical and academic work into accessible, policy-relevant conclusions. Small groups of professionals, meeting every few months, simply cannot provide quick turnaround in support of immediate needs. They cannot be a source of ongoing conversation and debate. They cannot help to vet the resources and research upon which a policymaker must base a decision. They were designed for an earlier, more settled era in which there was a belief in the "right" answer. That era has passed.

The institutions and tools of policymaking needed today must reach out more broadly, more collaboratively, in something closer to real time. They must permit—indeed, enable—not one-time spot solutions, but continuous experimentation and learning. Agencies do much more than enact rules. They have tremendous resources to bring to bear every day through their contracting, grantmaking, and convening power and therefore need ongoing "conversations" to be part and parcel of normal agency operations.

Moving toward this model demands ignoring the false dichotomy between reliance on a professional elite or on an ignorant citizenry, as well as the false choice between apolitical administration dominated by facts and democratic legislative practices based on values. Happily, when we take citizen expertise and the new opportunities to make use of it seriously, we can conceive of institutional arrangements that are more broadly participatory *and* more expert.

We can bring participatory practices to governance, including to administrative agencies, by expanding the role for citizen experts and diversifying the conception of engagement to include the daily work making policies and delivering services. Access to expertise is what matters. The need is not for unfettered access to more suggestions, more comments, more raw data, or more voices around the proverbial table, especially when they are partisan voices. Noise alone will not make it any easier for policymakers to decide or decide well. Instead, what is needed is ready access to the variety of expertise required, wherever it is located—from engineers and poets, chemists and historians, who, given the chance to self-select, would do so well and willingly—to gauge the validity and usefulness of the ideas and information already present within the torrent of available advice. That is the key.

The way to make government institutions more agile and effective is to reestablish them in constant conversation with an engaged, diverse, and knowledgeable public. Instead of having a discussion about tackling, say, the student debt load with just two dozen economists and university presidents, it becomes possible to look at the broader historical, social, and contextual environment in which the problem has arisen and ask questions of a much bigger network. If the topic is guns and violence in classrooms, policymakers can convene people with obviously relevant backgrounds in psychology and criminology, but also those with knowledge of gun manufacture, sensor technology, architecture, media studies, and local conditions, as well as practical experience surviving and aiding in a school shooting. It also becomes possible to sift and sort people to create groups that are intentionally diverse and that include students, teachers, and parents, but also venture capitalists and businesswomen. And it also becomes possible to organize multiple, parallel groups of people with common skills and backgrounds.

Of course, critics point to partisan braying and bickering—a world in which every day is Election Day—as evidence that the "dysfunction gauge has moved into the danger zone."[37] They are certainly right. Out-of-control partisanship does not serve society well. But they are also wrong. A polite and civil rearrangement of the deck chairs on the *Titanic* would have saved no one. Excessive partisanship, as unpleasant as it is, is a distraction from the real problem—and the real opportunity. We do not need to make old-style decisions in a kinder, gentler manner. We need to refashion how our institutions make decisions in the first place.

Opening up government need not ignore politics and partisanship. To the contrary, it is a *response* to them. It is hard to ignore facts in favor of partisan posturing when information is more visible and accessible. It is hard to defend poor, ill-informed choices when the relevant expertise is much easier to tap. It is hard to accept political wins that result in poor outcomes when intelligent experimentation shows more clearly what works well. And it is hard to be content with mediocre programs that persist because they are the pet project of a representative or senator when openness results in effective ways to work around them.

To make governance open, instead of relying on commissions and councils of elite and credentialed professionals, no matter how expert, we need ways to tap evolving expertise to respond and adapt to challenges as they change. We need real-time, agile, transdisciplinary advisory processes that recognize and make use of the broader spectrum of citizen expertise—inside government, credentialed and elite, and experience-based. We need not deny uncertainties but learn to live and work with them.[38]

Undertaking the challenge of moving systemically from insular and closed-door to open and participatory institutions of governing will require attacking the opportunity from many directions: the adoption of new technologies of expertise (Chapter 4), the development of an experimentation and research agenda (Chapter 5), the creation of supporting legal frameworks (Chapter 6), and the practical design of real-world implementations like the Brain Trust (Chapter 7).

The first step in getting from here to there is a clear and repeated articulation by world leaders, public intellectuals, activists, and bloggers of the core idea: the imperative need to take the capacity and expertise of citizens seriously and to put it to use in service of our democracy. To move from industrial to network-age designs for institutions, a frequent and crystalline pronouncement of the principles underlying this project will be essential to persuading people to undertake the necessary changes. Given how radical a departure these participatory ways of working are from the status quo, we cannot declare, define, and repeat often enough what it could mean to embrace collaboration and co-creation; to make consultation part of operations on a day-to-day basis; to strive for constant conversation with an engaged and knowledgeable public, and to reinvent the conception of public service and of the public servant as the steward of such a conversation.

Just as King Henry II invented the jury in the twelfth century and thus handed power to citizens in a practical and transformative fashion, we are at the threshold of being able to create these new institutional mechanisms. But that will require going beyond principles and pronouncements to create the platforms that make it possible. By analogy, look at what data.gov did to make the open data movement a reality.

President Obama's Memorandum on Transparency and Open Government, together with the subsequent G8 Charter on Open Data, put wind in the sails of those pushing for greater data transparency. But there had been freedom-of-information policies in prior years. What made this different was that the principle of open data was complemented by a platform: Data.gov was a tangible, concrete place for the public to find—and agencies to point to—their data.

Making progress against our goals to create "conversational" governing organizations will require experimenting with designs for networks like the Brain Trust. This is why there is a companion website to this book (Smarter State, http://www.smarterstate.org), designed to catalyze robust collaboration and conversation about mockups and prototypes for the right tools and processes to use at each stage of decisionmaking. To implement smarter government in practice, we need to show and convince ourselves through demonstration that it can be done. But to get at the "right horse for each course"—the appropriate choice of platform and practice for enhancing every stage of decisionmaking from problem spotting to solution design—requires deploying actual expert networks in public contexts and combining them with different crowdsourcing designs to see if and how targeting expertise works.

Creating platforms and integrating them into operations will require adopting a lean and agile mindset that embraces experimentation. If we are, for example, to get the Department of Education connected to and operating with the Brain Trust, there will have to be considerable trial and error. The department has to be willing to try and try again—do, learn, do—with various platforms and decision rules in order to come up with a consistent set of practices. Embracing experimentation, including randomized controlled trials, represents a radical shift from the world in which government must adopt the veneer of professional infallibility reflected in a uniform and unchanging set of practices.

Changing how we make decisions will depend squarely on having the personnel who embrace openness and collaboration. The recognition of citizen expertise does not mean jettisoning the professionals and substituting some kind of web-based plebiscite—far from it. Making good use of citizen expertise will require even better-trained managers and leaders. This new breed of professional—if we can call it that with a wink and a nod—will need to know how to work and talk and decide *with* citizens rather than for them. Just as the next generation will need to know how to govern with big and open data, predictive analytics, and real-time indicators, they will also need to develop the skillset for managing the conversation to get at the intelligence necessary to govern in a complex society. The new civil servant will be able to coordinate multiple channels for dialog, viewing these processes as core, and not incidental, to the job. The demand for leaders of such conversational organizations should create pressure for new curricula and training to meet the need.

But new principles, platforms, experimental practices, and training, will not work—and will not last—without empirical evidence. The loftiest words and the shiniest platforms do not create the impetus to undo a century and more of professional public management behind closed doors. Yes, for smarter governance to scale and deliver real value, compelling new ideas need to seize the imagination at the right moment in history. But good ideas get strength and staying power only when they are truly evidence-based and evidence-adjusted. There needs to be hard data showing that the technologies of expertise *can* make it possible to govern more effectively and *can* improve people's lives. Without it, any commitment to the effort will be short-lived.

There needs to be evidence that participatory governance works better. We have to go beyond faith to concrete, specific, and measurable strategies for making sure investments in platforms like the Brain Trust are merited. This demands an integrated framework for designing and implementing a rigorous research agenda to collect evidence on the impact of smarter governance, identifying specific results, and accounting for differences among populations and settings. What kinds of evidence can we gather to help measure the impact? What methodologies are most effective at gathering actionable evidence? How can we use evidence to differentiate and prioritize among initiatives? What steps are required to create an environment of constant learning?

Centralized government based on bureaucratic control enforced by law and reinforced by closed-door tradition and hierarchy is out of date. The open and networked governance institutions of the future will work smarter. They have to.

Conclusion: The Daedalus Project

> Thus it is that democracy, if it is to survive the shrinking of the world and the assaults of a hostile modernity, will have to re-discover its multiple voices and give citizens once again the power to speak, to decide, and to act; for in the end human freedom will not be found in the caverns of private solitude but in the noisy assemblies where women and men meet daily as citizens and discover in each other's talk the consolidation of a common humanity.
>
> —Benjamin Barber, *Strong Democracy*, 1984

Around the world, governments are faced with a challenge: deliver services, make policies and solve problems in ways that are more effective and transparent. Innovations such as open and big data, citizen science, and prize-backed public challenges to engage citizens—programmers, activists, journalists, teachers—are being presented as the first steps in helping governments become more open, more participatory, and more accountable. These more "conversational" approaches to administration are aimed at getting new opinions and ideas into government, but also at benefiting from the elbow grease and expertise of both citizens and civil servants.

Though making a commitment to innovation is the first step, achieving this goal requires going beyond thrusting the occasional reformer into an open government role, or recruiting the Silicon Valley techie to do a tour of duty in Washington or another capital. Improving the level of expertise in government requires more than tweeting a question or implementing a one-time open-call crowd-sourcing project. Instead we need to use new tools to unlock talent and systematically connect motivated innovators both inside and outside of government.

Today government gets expertise from many sources, both internal and external, but as we have discussed, it is often difficult to find the right expertise quickly enough and early enough. It is not enough to get comments after the fact in response to an already-drafted regulation, for example. Help at all stages of policymaking and service delivery is necessary if we are to be more innovative and efficient in how we govern. The growing demand for specific facts, data, and examples points, as well, to a parallel need to diversify sources and types of expertise. More people with relevant knowledge must be able to self-select to participate in a "community" from which they can be easily called upon by government to work on important issues. As we have seen, expert networking tools are already in use to help academics find one another to undertake research collaborations. Imagine applying this same kind of expert networking to enable those inside and outside the academy to collaborate on tackling significant public problems together.

The Internet is radically decreasing the costs of identifying diverse forms of expertise and segmenting audiences on the basis of credentials, experiences, skills, and interests. New technologies of expertise, like expert networks, are multiplying the number and type of expertise, including skills and experiences, that can be systematically searched. Because expertise can now be identified and collected both manually and automatically, and can be calculated based on different traits and characteristics, it is possible to quantify expertise differently and more diversely than before. By divorcing the concept of expertise from elite social institutions and creating tools to enable neutral identification of talent and ability—whether of those inside or outside of government, with credentials or craft knowledge—technology is democratizing expertise for the subject being described and especially for the institutions trying to search for it.

The transformation of expertise is playing out against the backdrop of broader developments in education and society. At the end of the nineteenth century, the rise of the research university was precipitated by and also catalyzed the rise of professions. Today, the unbundling of education and the distribution of control over education to new online platforms may be dismantling the domination of social life by professions (and the high value of the credential) in the twenty-first century with far-reaching implications for governance. With the ability to decentralize how we demonstrate and certify expertise and mastery, we are concomitantly making expertise more searchable and usable. The two trends are inextricably linked because of the central role of training and education in the ideology of professions.

In this "maker generation" of 3D printing and popular mechanics for the digital age, there is a renewed interest in craft know-how and learning-by-doing and ideation rather than rote learning. There is a celebration of creativity over memorization. The White House held a National Day of Making in June 2014 to celebrate the American inventive spirit and to encourage the development of new maker spaces and investment in mentoring young people through hands-on learning. MIT, probably the nation's preeminent science university, now provides an official mechanism for applicants to share a maker portfolio of what they have made and created outside of school.[1] Countless pedagogues are championing the wisdom of play and personal creativity as a pathway to learning.[2] In an interview with Thomas L. Friedman, Laszlo Bock, the senior vice president of people operations for Google is blunt on this point: "G.P.A.'s are worthless as a criteria for hiring, and test scores are worthless. . . . We found that they don't predict anything." He also notes that the "'proportion of people without any college education at Google has increased over time'—now as high as 14 percent on some teams."[3] The least important attribute Google looks for in a candidate is expertise—that is, expertise understood traditionally as brand-name university training. Bock's reaction is consistent with an increasingly widespread view that universities are overcharging and, at the same time, failing to turn out graduates with skills and abilities.

Of course, the growth of universities in the United States also had a democratizing effect on the economy and society. Proportionately, more people attend college in this country than in any other. This has had the effect of boosting rates of primary school and secondary school education among all classes of society. Now, just after the turn of the twenty-first century, rising inequality, lagging economic productivity, and a failing report card for the country's educational system are calling into question the quality of the structure we have created. "School" is broken, and everyone knows it.[4]

For too many people, higher education is leading to the demise, not the enrichment, of economic opportunity.[5] "If you want to know what college is actually like in this country," writes Clay Shirky, "forget Swarthmore, with 1500 students. Think Houston Community College, with 63,000. Think rolling admissions. Think commuter school. Think older. Think poorer. Think child-rearing, part-time, night class. Think fifty percent dropout rates. Think two-year degree. (Except don't call it that, because most graduates take longer than two years to complete it. If they complete it.)" Think internships provided by others, not instruction provided at school. According to Shirky, "The end game is degrees that are little more than receipts for work done elsewhere. Empire State, Excelsior, Thomas Edison, all these institutions and more convert a loose set of credits into a diploma, without much of anything resembling a curriculum."[6] Traditional college is getting worse, not better, for the majority of Americans. The failure of our education system to deliver value in the current economic climate is in the news daily.

But from the ashes of this crisis within the institutions of higher education the phoenix of great opportunity is rising. Even as the traditional education system is in crisis, educational attainment rates are at their highest levels in history. Since the birth of the Internet and the commercial World Wide Web, there has also been an explosion in amateur content production. "Following in the footsteps of pioneers like Wikipedia and Linux, again and again the creation of new, free or inexpensive, easy to use platforms released waves of human creativity, entrepreneurial aspiration and collaborative endeavor," says Matthew Taylor, chief executive of the RSA (Royal Society for the Encouragement of Arts, Manufactures and Commerce) and one of innumerable essayists to marvel at the age of creativity in which we live.[7]

Marry the dissatisfaction with current educational models to new platforms for expressing expertise—from LinkedIn to Stack Exchange to Khan Academy, which provides detailed analytics of how a student learns—and making new kinds of expertise systematically discoverable and it becomes possible to demand experience over credentials, certified skills over mere certifications, and practical ability over status hierarchy. These platforms incorporate new features for demonstrating mastery and for allowing others to rate and recommend those who possess it. Whether in computer games like *World of Warcraft* or commercial marketplaces like EBay and Amazon, participants identify who are the builders or warriors or power sellers and target those with whom they wish to play, do business, and create. As a result, people now regularly collaborate in constructing worlds, making markets, and creating art in ways that far outstrip anything that a professional designer could do alone.

When we can see with precision who knows what, we can harness that know-how for the public good—that is, make our democracy, starting with the administrative state, which manages the day-to-day affairs of governing, more participatory. The myth of "a few good men," where the best and brightest technocrats could be relied on to govern well because of their university training, seems quaint. It is possible finally to realize what philosopher Danielle Allen describes as the "egalitarianism of co-creation and co-ownership of a shared world, an expectation for inclusive participation that fosters in each citizen the self-understanding that she, too, he, too, helps to make, and is responsible for, this world in which we live together."[8]

The technologies of expertise make it possible to go beyond the proxies of credentials and professional membership and imagine more agile, diverse, ongoing communities of expertise like the Brain Trust. Tested platforms for collaboration and crowdsourcing can then support both large- and small-scale convenings, long- and short- term interventions, and a whole range of mechanisms—both task-based and conversational—as yet untried in the public sector for infusing decisionmaking with expertise and, at the same time, infusing citizenship with substantive opportunities for participation.

That much is now clear. Also clear, however, is the fact that this revolution in participatory culture has not and will not come to government of its own accord. All too often ignorance about politics is still considered ignorance. People may not be conversant in the sport of politics, but they do possess expertise, understood broadly, in spades. It is incumbent upon those who govern to get at that know-how, not occasionally, but continuously. Yet for too long, as a result of history, theory, and institutional practice, citizens have been largely excluded from governing. Our institutions are not designed to allow, let alone encourage, rich, ongoing collaboration.

The impediments to much-needed institutional redesign and innovation are not limited to legal safeguards against nepotistic behavior. Even if these legal challenges were circumvented or removed, there is no value in plugging citizen expertise into structures not designed to use it. Before they can take advantage of such expertise, our institutions must be configured to use it well and consistently. Today, they are not. Typically, they use it for specific, bounded transactions, not as a regular part of an ongoing conversation between state and citizen.

Modern communications technology, growing experience with organizing collaborative activity at a distance, and recent policy changes together create the prospect for practical change. Many examples already exist of the state taking advantage of citizen expertise, especially programming and technology know-how, through assorted crowdsourcing activities, including challenges and hackathons. These "open calls" are inspiring when they work, but they are too random to be the basis for sustained organizational innovation. They do not ensure that a large enough or diverse enough mix of people will come together regularly to justify institutional redesign. Merely layering on citizen deliberation exercises perpetuates the myth of citizen ignorance, deprives the polity of much-needed problem-solving ability, and justifies continued reliance on professionalized closed-door processes for governing.

The broken, staccato rhythm of citizen engagement today does not, and cannot, support meaningful action on the complex problems we face—problems that cannot be solved, only tackled, and then re-tackled as they evolve. For real change, institutions of governance must take consistent, regular advantage of the best that people inside and outside of government have to offer. Already, of course, as voters and jury members, citizens do participate, but only in limited ways that use a small portion of their capacity. The requirements to vote and serve are very low—not in the amount of time, but in the quality and nature of engagement. And when people are infrequently asked to do more, to participate in deliberative polls or citizen assemblies, they are asked to watch, but not act, to keep government accountable but not to join in, to be educated on issues but not to contribute.

Social media, which channel a torrent of self-expression from a wide range of participants, are largely disconnected from formal channels of modern decisionmaking. Advisory groups, committees, and panels do rely on the expertise of those who serve, but they usually draw only from the most-credentialed elites to develop a report or set of recommendations on a specific course of policy action. They are still mired in the politics of interest group and stakeholder representation. For all practical purposes, the closed door of professional governance remains closed.

This is no longer a tolerable state of affairs.[9] We urgently need to take affirmative steps to design institutions to take advantage of citizen expertise and creativity for the public interest—more *Minecraft*, less statecraft. Amazingly, however, although there has been a mountain of writing about the relationship between expertise and politics, the question of the *appropriate institutional design* for regularly incorporating citizen expertise into the normal, ongoing conversations of public decisionmaking is largely unexplored.[10] That is the goal of this book.

There is no more important public issue today than how to develop our governing organizations to make them smarter and better able to tackle the myriad and complex challenges we face. We can and will continue to debate which policy is better and more likely to lead to a desired outcome. But this is of secondary importance and urgency to reinventing the processes by which we make policy in the first place. Over the next fifty years, we will face challenges that no previous generation of humanity has ever had to deal with. To overcome, we have to run our communities and our institutions differently. The crisis of human capital, to paraphrase Sir Ken Robinson, is our other climate crisis—making real use of what people know.

The very real opportunity that is upon us is to innovate in how we get the full gamut of this expertise—formal scientific knowledge and credentials, but also skills and experiences—into policymaking. To take advantage of the revolution in expertise technologies will require crossing academic and disciplinary chasms to foster broader debate about and experimentation with new institutional designs. What we need is a Daedalus project for the twenty-first century.

Daedalus, according to the myth, was the first science and technology adviser to design clever solutions for a king's problems—a wooden cow to conceal the king's wife as she mated with a bull; a labyrinth to contain the fearsome minotaur; and of course, the wax-reinforced artificial wings that he planned to use to fly home from exile and that his son, Icarus, used to fly too close to the sun, instead.

None of his inventions was perfect. None worked well forever. All failed at some point. No one invention or answer was ever permanently right. When a solution wore out or miscarried, whether from human or mechanical failure, Daedalus devised another. He was a skilled craftsman and maker. Naturally, the stories about him highlight his vast expertise; more than that, they highlight his capacity to change and adapt his solutions as circumstances changed around him. For us, today, he represents the power, not of pure science or raw data but of practical innovation and finding more effective ways of tackling ever-shifting challenges in a complex environment. Building and maintaining the institutions and processes to sustain a two-way conversation between government and large numbers of citizens—not uniform, one-size-fits-all but diverse opportunities that speak to people's talents, abilities and interests—is therefore the ideal Daedalus project for our time. This is *our* design challenge.

Notes

1. From Open Government to Smarter Governance

1. Dorothy Nelkin, "The Political Impact of Technical Expertise," *Social Studies of Science* 5, no. 1 (1975): 35–54.
2. Rebecca Rifkin, "Public Faith in Congress Falls Again, Hits Historic Low," *Gallup Politics*, June 19, 2014, http://www.gallup.com/poll/171710 /public-faith-congress-falls-again-hits-historic-low.aspx.
3. "Government by and for Millennial America," Roosevelt Campus Network, July 2013, http://gbaf.tumblr.com/post/42945987007/while -millennials-strongly-believe-in-an-activist.
4. Edelman, "2014 Edelman Trust Barometer Global Results," 2014, http://www.edelman.com/insights/intellectual-property/2014 -edelman-trust-barometer/about-trust/global-results/.
5. Henry Chesbrough, "Open Innovation: A New Paradigm for Understanding Industrial Innovation," in *Open Innovation: Researching a New Paradigm,* ed. Henry Chesbrough, Wim Van-haverbeke, and Joel West, (New York: Oxford University Press, 2006), 1–12.
6. See, e.g., Karim R. Lakhani, Hila Lifshitz-Assaf, and Michael Tushman, "Open Innovation and Organizational Boundaries: Task Decomposition, Knowledge Distribution, and the Locus of Innova-tion," in *Handbook of Economic Organization: Integrating Economic and Organizational Theory,* ed., Anna Grandori (Cheltenham, UK: Edward Elgar, 2013), 355–82; see also Frank Piller, Christoph Ihl, and Alexander Vossen, "A Typology of Customer Co-creation in the Innovation Process," SSRN, December 29, 2010, http://papers.ssrn .com/sol3/papers.cfm?abstract_id=1732127.

7. "My Starbucks Idea," Starbucks, 2010, http://www.starbucks.com /coffeehouse/learn-more/my-starbucks-idea.

8. "IdeaStorm Can Help Take Your Idea and Turn It into Reality," Dell, 2015, http://www.ideastorm.com/.

9. "Netflix Prize," Netflix, 2009, http://www.netflixprize.com//commu nity/viewtopic.php?id=1537.

10. Facebook, "Investor Relations: Facebook Reports First Quarter 2013 Results," May 1, 2013, http://investor.fb.com/releasedetail.cfm ?ReleaseID=761090. Facebook Enterprise Value, 84.84B; Viacom Enterprise Value, 46.13B; CBS Enterprise Value, 38.53B. Y Charts, August 12, 2013.

11. See, e.g., Risto Rajala et al., "From Idea Crowdsourcing to Managing User Knowledge," *Technology Innovation Management Review,* December 2013, http://timreview.ca/article/750.

12. See, e.g., Nick Van Breda and Jan Spruijt, "The Future of Co-Creation and Crowdsourcing," Edcom Annual Conference, 2013, http://papers .ssrn.com/sol3/papers.cfm?abstract_id=2286076.

13. The Governance Lab, "The Story of the Open Source Drug Discovery Project: Dr. Samir Brahmachari," *YouTube,* May 1, 2015, https://www .youtube.com/watch?v=WdGzbGQHV8U. See also Vegard Kolbjorn- srud, "On Governance in Collaborative Communities, Norwegian Business School" (PhD diss., Business School, University of Norway, 2014), 61–64.

14. See, e.g., Tanja Aitamurto, "The Promise of Idea Crowdsourcing: Benefits, Contexts, Limitations," http://walsww.academia.edu/963662 /The_Promise_of_Idea_Crowdsourcing_Benefits_Contexts_Limita- tions. Thomas W. Malone, Robert Laubacher, and Chrysanthos Dellarocas, "Harnessing Crowds: Mapping the Genome of Collective Intelligence" (working paper, MIT Center for Collective Intelligence Working Paper, Cambridge, MA, February 2009), http://cci.mit.edu /publications/CCIwp2009-01.pdf. Eric Bonabeau, "Decisions 2.0: The Power of Collective Intelligence," *MIT Sloan Management Review* (Winter 2009): 45–52. Benjamin Y. Clark et al., "A Framework for Using Crowdsourcing in Government," SSRN, July 8, 2013,, http://ssrn .com/abstract=1868283 or http://dx.doi.org/10.2139/ssrn.1868283.

15. Friedrich Hayek, "The Use of Knowledge in Society," *American Economic Review* 35, no. 4 (1945): 519–30.

16. For a raft of such examples, see Steven Berlin Johnson, *Future Perfect: The Case for Progress in a Networked Age* (New York: Riverhead, 2012).

17. Jim Yardley, "With Survey, Vatican Seeks Laity Comment on Family Issues," *New York Times,* November 8, 2013, http://nyti.ms/1adv6Jl. For example, Question 1a Describe how the Catholic Church's teachings on the value of the family contained in the Bible, Gaudium et Spes, Familiaris Consortio and other documents of the post-conciliar Magisterium is understood by people today? What formation is given to our people on the Church's teaching on family life? See https://www.surveymonkey.com/s/FamilySynod2014.

18. Opening Remarks by President Obama on Open Government Partnership (September 20, 2011), http://www.whitehouse.gov/the-press-office/2011/09/20/opening-remarks-president-obama-open-government-partnership.

19. Written Declaration pursuant to Rule 123 of the Rules of Procedure on open and collaborative government (May 9, 2012), http://www.europarl.europa.eu/sides/getDoc.do?pubRef=-//EP//NONSGML+WDECL+P7-DCL-2012–0019+0+DOC+PDF+V0//EN&language=EN.

20. Prime Minister David Cameron, Speech at Open Government Partnership Summit (October 31, 2013), https://www.gov.uk/government/speeches/pm-speech-at-open-government-partnership-2013.

21. Moises Naïm, "Why the People in Power Are Increasingly Powerless," *Washington Post,* March 1, 2013, http://articles.washingtonpost.com/2013–03–01/opinions/37367377_1_climate-talks-president-obama-power/3.

22. See also Tanja Aitamurto, "The Promise of Idea Crowdsourcing: Benefits, Contexts, Limitations," http://www.academia.edu/963662/The_Promise_of_Idea_Crowdsourcing_Benefits_Contexts_Limitations.

23. Sarah Rosenberg, "Restoring Pell Grants to Prisoners: Great Policy, Bad Politics" (InformEd blog), *American Institutes for Research,* November 5, 2012, http://www.quickanded.com/2012/11/restoring-pell-grants-to-prisoners-great-policy-bad-politics.html.

24. Lois M. Davis, Robert Bozick, Jennifer L. Steele, Jessica Saunders, and Jeremy N. V. Miles, *Evaluating the Effectiveness of Correctional Education: A Meta-Analysis of Programs That Provide Education to Incarcerated Adults,* RAND Report 2013.

25. C. Henrichson and R. Delaney, "The Price of Prisons: What Incarceration Costs Taxpayers," Center on Sentencing and Corrections, January 2012, http://www.vera.org/pubs/special/price-prisons-what-incarceration-costs-taxpayers; Partakers, "College behind Bars, http://partakers.org/site/college-behind-bars/.

26. Dustin Haisler, interview with the author, September 16, 2013.

27. Mazdoor Kisan Shakthi Sangathan (MKSS India), http://www.mkssindia.org/.

28. Redbridge Online Budget Simulator, "YouChoose," http://youchoose.esd.org.uk/redbridge2012.

29. Kaggle website, http://www.kaggle.com, and DrivenData website, http://www.drivendata.org/.

30. See, e.g., Maged N. Kamel Boulos et al., "Crowdsourcing, Citizen Sensing and Sensor Web Technologies for Public and Environmental Health Surveillance and Crisis Management: Trends, OGC Standards and Application Examples," *Internal Journal of Health Geographics* 10, no. 67 (2011), http://www.biomedcentral.com/content/pdf/1476–072X-10–67.pdf.

31. Jessica McKenzie, "Can Patrick Meier's New App MicroMappers Completely Change the Way We Think about Clicktivism?" *TechPresident,* September 19, 2013, http://techpresident.com/news/wegov/24356/can-patrick-meiers-new-app-micromappers-completely-change-way-we-think-about-clicktivism.

32. Honduras Health Mapping, https://hondurashealthmapping.crowdmap.com/main; Mapa Delectivo, http://www.eluniversal.com.mx/graficos/oocoberturas/mapa_delictivo/.

33. Jan Behrens et al., *The Principles of Liquid Feedback* (Berlin: Interactive Demokratie, 2014).

34. Hazel Markus, "The Effect of Mere Presence on Social Facilitation: An Unobtrusive Test," *Journal of Experimental Social Psychology* 14 (1978): 389–97. The principles of the audience effect are, in part, what underlie the idea of the citizen jury or the citizen panel, namely the convening of a small group of citizens to provide input or exercise oversight over an official process. See The Jefferson Center, "Citizen Jury," http://jefferson-center.org/how-we-work/citizen-juries/.

35. Tetherless World Constellation Web Observatory Portal, Rennselaer Polytechnic Institute (2014), http://tw.rpi.edu/web/web_observatory.

36. Tom Lee, "Defending the Big Tent: Open Data, Inclusivity, and Activism," *The Sunlight Foundation Blog,* May 2, 2012, http:// sunlightfoundation.com/blog/2012/05/02/defending-the-big-tent -open-data-inclusivity-and-activism/; Ethics.Data.gov, http://www .ethics.gov/; Visitor Access Records, The White House Briefing Room, February 27, 2015, http://www.whitehouse.gov/briefing-room /disclosures/visitor-records; USAspending.gov, http://www .usaspending.gov/.

37. Directive on Open Government, Treasury Board of Canada Secretariat, October 9, 2014, http://www.tbs-sct.gc.ca/pol/doc-eng.aspx?id =28108.

38. Peter R. Orszag, "Memorandum for the Heads of Executive Departments and Agencies, Executive Office of the President of the United States," Office of Management and Budget, December 8, 2009, http://www.whitehouse.gov/omb/assets/memoranda_2010 /m10-06.pdf.

39. James Manyika, Michael Chui, Diana Farrell, Steve Van Kuiken, Peter Groves, and Elizabeth Almasi Doshi, "Open Data: Unlocking Innovation and Performance with Liquid Information," McKinsey & Company, October 2013, http://www.mckinsey.com/insights/business _technology/open_data_unlocking_innovation_and_performance _with_liquid_information.

40. Beth Simone Noveck and Daniel L. Groff, *Information for Impact: Liberating Nonprofit Sector Data* (Washington, DC: The Aspen Institute, 2013), http://www.aspeninstitute.org/sites/default/files /content/docs/psi/psi_Information-for-Impact.pdf.

41. See "Full Text of the Magna Carta for Philippine Internet Freedom (MCPIF)," http://democracy.net.ph/full-text/; The Open Ministry in Finland, for example, is a nonprofit organization that promotes "crowdsourcing legislation, deliberative and participatory democracy and citizens initiatives" by facilitating citizen debate over and discussion of draft legislative proposals for submission to the Finnish Parliament. See "Crowdsourcing Legislation," http://openministry .info/. The Particepa! project on constitutional reform in Italy ran a three-layer, sequential public consultation initiative in 2013 to legitimize constitutional reform efforts by fostering new channels for greater and more meaningful citizen input into the process. See

http://partecipa.gov.it/. See also "Consultazione Pubblica sulle
Riforme Costituzionali Rapporto Finale," http://www.riforme
costituzionali.partecipa.gov.it/. See the e-Democracia platform used
within the Brazilian House of Representatives, http://edemocracia
.camara.gov.br/ and participa.br, developed and run by the
General Secretariat of the Brazilian Presidency, a ministry responsible
for social participation and social movements, http://www.participa
.br/.df. See Andrew Mandelbaum, "Online Tools for Engaging
Citizens in the Legislative Process," OpenParliament.org., February 28, 2014. Jillian Raines, "Toward More Inclusive Lawmaking:
What We Know and Still Most Need to Know about Crowdlaw"
(blog), *The GovLab,* June 4, 2014, http://thegovlab.org/toward-more
-inclusive-lawmaking-what-we-know-still-most-need-to-know-about
-crowdlaw.

42. "Icelandic Constitutional Council 2011," *Participedia,* http://
participedia.net/en/cases/icelandic-constitutional-council-2011; Sean
Deely and Tarik Nesh Nash, "The Future of Democratic Participation:
my.con: An Online Constitution Making Platform," http://ceur-ws
.org/Vol-1148/paper4.pdf.

43. Nathan Schneider, "Some Assembly Required: Witnessing the Birth
of Occupy Wall Street," *Harper's,* February 2012.

44. Manuel Castells, *Networks of Outrage and Hope: Social Movements in
the Internet Age* (New York: Polity Press, 2012).

45. Jeff Jarvis, "#OccupyWallStreet and the Failure of Institutions,"
BuzzMachine, October 3, 2011, http://buzzmachine.com/2011/10/03
/occupywallstreet-the-failure-of-institutions/.

46. "Gladwell vs. Shirky: A Year Later, Scoring the Debate over Social-
Media Revolutions | Threat Level," *WIRED,* December 27, 2011,
http://www.wired.com/2011/12/gladwell-vs-shirky/.

47. Zeynep Tufekci, "After the Protests," *New York Times,* March 19, 2014,
http://www.nytimes.com/2014/03/20/opinion/after-the-protests.html.

48. Susan L. Moffitt, *Making Policy Public: Participatory Bureaucracy in
American Democracy* (Cambridge, UK: Cambridge University Press,
2014).

49. Francis Fukuyama, "The Decay of American Political Institutions,"
American Interest, December 8, 2013, http://www.the-american-interest
.com/articles/2013/12/08/the-decay-of-american-political-institutions.

50. Alice Lipowicz, "Use Digital Tools for Better E-rulemaking, Former Official Advises," *FCW,* January 26, 2011, http://fcw.com/articles/2011 /01/26/former-white-house-deputy-cto-advises-immediate-actions-for -improved-erulemaking.aspx.

51. See Robert Steinbrook, "Science, Politics, and Federal Advisory Committees," *New England Journal of Medicine* 350, no. 14 (April 2004): 1454–1460.

52. See Robin M. Nazzaro, "Issues Related to the Independence and Balance of Advisory Committees," Highlights of Testimony before the Subcommittee on Information Policy, Census, and National Archives, Committee on Oversight and Government Reform, House of Representatives, April 2, 2008, http://www.gao.gov/assets/120 /119486.pdf.

53. See Beth Simone Noveck, *Wiki Government: How Technology Can Make Government Better, Democracy Stronger, and Citizens More Powerful* (Washington, DC: Brookings Institution Press, 2009), 131.

54. Andrew Rich, *Think Tanks, Public Policy and the Politics of Expertise* (Cambridge, UK: Cambridge University Press, 2004), 34.

55. Ibid., 28.

56. Eric Lipton, Brooke Williams, and Nicholas Confessore, "Foreign Powers Buy Influence at Think Tanks," September 6, 2014, http://www .nytimes.com/2014/09/07/us/politics/foreign-powers-buy-influence-at -think-tanks.html?_r=0.

57. Peer to Patent is the central case study in Noveck, *Wiki Government.*

58. Marisa Peacock, "The Search for Expert Knowledge Continues," *CMSWire,* May 12, 2009, http://www.cmswire.com/cms/enterprise -cms/the-search-for-expert-knowledge-continues-004594.php.

59. Clive Thompson, *Smarter Than You Think: How Technology Is Changing Our Minds for the Better* (New York: Penguin Press, 2013), 62.

60. Donald McNeil Jr., "Car Mechanic Dreams Up a Tool to Ease Births," *New York Times,* November 13, 2013, http://nyti.ms/17UuZos.

61. " 'And the Winner Is . . .' Capturing the Promise of Philanthropic Prizes," McKinsey & Company, July 2009, http://mckinseyonsociety .com/downloads/reports/Social-Innovation/And_the_winner_is.pdf.

62. "Implementation of Federal Prize Authority: Fiscal Year 2012 Progress Report," Office of Science and Technology Policy, The White House,

December 2013, http://www.whitehouse.gov/sites/default/files
/microsites/ostp/competes_prizesreport_dec-2013.pdf. See also
"Prize-Backed Challenges" (wiki), *The GovLab,* http://thegovlab.org
/wiki/Prize-backed_Challenges.

63. Cristin Dorgelo and Tom Kalil, "21st Century Grand Challenges"
(blog), *White House,* April 9, 2012, http://www.whitehouse.gov/blog
/2012/04/09/21st-century-grand-challenges.

64. *America Competes Act,* Public Law 110–69, August 9, 2007.

65. Case Foundation, "Collaborative Innovation: Dr. Alok Das and Jean
Lupinacci," *YouTube,* August 21, 2012, http://www.youtube.com/watch
?v=dA3l_UlFZac. See also "InnoCentive and the Air Force Research
Lab Announce Successful Completion of Initial Open Innovation
Challenges," *InnoCentive,* October 17, 2011, http://www.innocentive
.com/innocentive-and-air-force-research-lab-announce-successful
-completion-initial-open-innovation-challe.

66. Jeff Howe, "The Rise of Crowdsourcing," *Wired* 14, no. 6 (2006): 1–4.

67. Frank Fischer, *Democracy and Expertise: Reorienting Policy Inquiry*
(Princeton, NJ: Princeton University Press, 2009), Kindle ed. location
59–60.

68. Robert Dahl, *Democracy and Its Critics* (New Haven, CT: Yale
University Press, 1989), 339.

69. Michael Schudson, *The Good Citizen: A History of American Civic Life*
(New York: Free Press, 2010), 23ff; See also Jeffrey Green, *The Eyes of
the People: Democracy in an Age of Spectatorship* (New York: Oxford
University Press, 2010).

70. Tim Causer, "Welcome to Transcribe Bentham," *Transcribe Bentham:
A Participatory Initiative,* March 27, 2013, http://blogs.ucl.ac.uk
/transcribe-bentham/. "Ancient Lives Project," *Ancient Lives,* http://
ancientlives.org/about.

71. Bill and Melinda Gates, "A Call for Global Citizens," *GatesNotes,*
http://www.gatesnotes.com/2015-annual-letter?WT.mc_id=01_21
_2015_DO_GFO_domain_0_00&page=0&lang=en. See also Bill
Gates, "How to Fight the Next Epidemic: The Ebola Crisis Was
Terrible. But Next Time Could Be Much Worse," *New York Times,*
March 18, 2015, http://www.nytimes.com/2015/03/18/opinion/bill-gates
-the-ebola-crisis-was-terrible-but-next-time-could-be-much-worse
.html?ref=opinion&_r=1.

72. Douglas W. Allen, *Institutional Revolution: Measurement and the Economic Emergence of the Modern World* (Chicago: University of Chicago Press, 2011).

73. Geoff Mulgan, "Systems, Success, Failure, and Rebirth," *Global Brief,* October 4, 2012, http://globalbrief.ca/blog/2012/10/04/systems-success -failure-rebirth/.

74. A case in point was the *Citizens Briefing Book* crowdsourcing exercise conducted during the 2008–2009 presidential transition. The transition team asked the American people to propose issues to be added to the president's first-100-day agenda. The original idea was to elicit novel proposals that could be incorporated into the transition team's briefing books. Rolled out only a few days before the inauguration, the results "lacked novel thinking" or useful proposals. James E. Katz, Michael Barris, and Anshul Jain, *The Social Media President: Barack Obama and the Politics of Digital Engagement* (New York: Palgrave Macmillan, 2013), 47.

75. Katherine Bourzac, "Biodegradable Batteries to Power Smart Medical Devices:Prototype Batteries that Dissolve Safely in the Body Could Power Ingested Devices," *MIT Technology Review*, December 18, 2013, http://www.technologyreview.com/news/522581 /biodegradable-batteries-to-power-smart-medical-devices/. See also, The Bettinger Research Group, Carnegie Mellon University, http://biomicrosystems.net/.

76. "Research on the Use of Science in Policy: A Framework," in *Using Science as Evidence in Public Policy*, ed. Kenneth Prewitt, Thomas A. Schwandt, and Miron L. Straf (Washington, DC: National Academies Press, 2012), http://www.nap.edu/openbook.php?record_id=13460 &page=53.

77. William Easterly, in *The Tyranny of Experts* (New York: Basic Books, 2014), makes the point that it is often experts who are most likely to trample the rights of the already downtrodden.

78. Atul Gawande, "Testing, Testing: The Health-Care Bill Has No Master Plan for curbing Costs. Is That a Bad Thing?" *New Yorker,* December 2009, 34–41.

79. BBC "Study: US Is an Oligarchy, Not a Democracy" (blog), *Echo Chambers,* April 17, 2014, http://www.bbc.com/news/blogs -echochambers-27074746.

80. Martin Gilens and Benjamin I. Page, "Testing Theories of American Politics: Elites, Interest Groups, and Average Citizens," *Perspectives on Politics* 12, no. 3 (2014): 564–81, http://dx.doi.org/10.1017 /S1537592714001595. John Cassidy, "Is America an Oligarchy?" (blog), *New Yorker*, April 18, 2014, http://www.newyorker.com/online/blogs /johncassidy/2014/04/is-america-an-oligarchy.html.

81. Michael Flowers, "Beyond Open Data: The Data-Driven City," in *Beyond Transparency: Open Data and the Future of Civic Innovation,* ed. Brett Goldstein and Lauren Dyson (San Francisco, CA: Code for America Press, 2013), 185–98. See also "Mayor Bloomberg and Speaker Quinn Announce New Approach to Target Most Dangerous Illegally Converted Apartments," press release (PR- 193–11), City of New York, June 7, 2011, http://www.nyc.gov/cgi-bin/misc/pfprinter.cgi?action =print&sitename=OM&p=1390075778000.

82. Geoff Mulgan, with Simon Tucker, Rushinara Ali, and Ben Sanders, "Social Innovation: What It Is, Why It Matters, How It Can Be Accelerated" (Oxford, UK: Skoll Centre for Social Entrepreneurship, University of Oxford, 2007), 5.

83. C. H. Weiss, "Improving the Linkage between Social Research and Public Policy," in *Knowledge and Policy: The Uncertain Connection,* ed. L. E. Lynn (Washington, DC: National Academies Press, 1978).

84. Mikey Dickerson, "Mikey Dickerson to SXSW: Why We Need You in Government," *Medium*, March 26, 2015, https://medium.com/@ USDigitalService/mikey-dickerson-to-sxsw-why-we-need-you-in -government-f31dab3263a0.

85. Source Document for Personnel Data, *New York City Police Depart- ment Form PD 446-141* (Rev. 11-14).

2. The Rise of Professional Government

Epigraph: George Bernard Shaw, *The Doctor's Dilemma* (New York: Penguin, 1946); the play was first performed in 1906.

1. Jerry L. Mashaw, *Creating the Administrative Constitution: The Lost One Hundred Years of American Administrative Law* (New Haven, CT: Yale University Press, 2012).

2. Bernard Crick, *The American Science of Politics: Its Origins and Conditions* (New York: Routledge, 1959). See also Bliss Perry, *The American Mind and American Idealism* (New York: Houghton Mifflin, 1913).

3. For more on the notion of reviving a citizens' literature, see Danielle Allen, *Our Declaration: A Reading of the Declaration of Independence in Defense of Equality* (New York: Liveright, 2014).

4. Crick, *American Science of Politics,* 9.

5. Stanley Elkins and Eric McKitrick, "A Meaning for Turner's Frontier: Democracy in the Old Northwest," *Political Science Quarterly* 69, no. 3 (1954): 336.

6. Richard Hofstadter, *The Age of Reform* (New York: Vintage Books, 1955), 133.

7. See Congressional Research Service, "History of Civil Service Merit Systems," 1976, http://archive.org/stream/hicivilseoolibr /hicivilseoolibr_djvu.txt; and Bureau of Labor Statistics, Occupational Employment Statistics, 2014, http://www.bls.gov/oes/current /naics4_999100.htm#00-0000.

8. Theodore Roosevelt, *The Works of Theodore Roosevelt: American Ideals and Other Essays* (New York: Scribner, 1906), 145.

9. The concept of profession is, "intrinsically ambiguous, multifaceted folk concept, of which no single definition and no attempt at isolating its essence will ever be generally persuasive," writes Eliot Freidson in *The Sociology of the Professions: Lawyers, Doctors and Others,* ed. Robert Dingwall and Philip Lewis (New Orleans, LA: Quid Pro Books, 2014), 32. For more discussion about the definitional conundrum, see Andrew Abbott, *The System of Professions: An Essay on the Division of Expert Labor* (Chicago: University of Chicago Press, 1988); Howard S. Becker, "The Nature of a Profession" in *Sociological Work* (Chicago: Aldine, 1970); Robert K. Merton, *Some Thoughts on the Professions in American Society* (Providence, RI: Brown University Papers 37, 1960); Robert K. Merton, *Social Theory and Social Structure,* 2nd rev. ed. (1949; New York: Free Press, 1968) Talcott Parsons, "Professions," in *International Encyclopedia of the Social Sciences,* vol. 12, ed. D. Sills and R. Merton (New York: Macmillan and The Free Press, 1968), 536–47.

10. One notable exception among political scientists is Frank Fischer, *Democracy and Expertise: Reorienting Policy Inquiry* (Princeton, NJ: Princeton University Press, 2009). For more on the literature of expertise, see also Dingwall and Lewis, *Sociology of the Professions;* and K. Anders Ericsson et al., eds., *The Cambridge Handbook of Expertise and Expert Performance* (Cambridge, UK: Cambridge University Press, 2006).

11. Dietrich Rueschmeyer, "Professional Autonomy and the Social Control of Expertise," in Dingwall and Lewis, *Sociology of the Professions,* chap. 2.

12. Paul Starr, *The Social Transformation of American Medicine: The Rise of a Sovereign Profession and the Making of a Vast Industry* (New York: Basic Books, 1984), 4.

13. Abbott, *System of Professions.*

14. In *Stuart v. Loomis,* 992 F. Supp. 2d 585, 588 (M.D.N.C. 2014), the federal district court held that the state does not have the power to compel a health care provider, who is governed by the rules of the medical profession, to deliver the state's ideological message in favor of carrying a pregnancy to term.

15. C. L. Gilb, *Hidden Hierarchies: The Professions and Government* (New York: Harper and Row, 1966).

16. Robert C. Post, *Democracy, Expertise and Academic Freedom: A First Amendment Jurisprudence for the Modern State* (New Haven, CT: Yale University Press, 2012), 95. See also Alexander Meiklejohn, "The First Amendment Is an Absolute," *Supreme Court Review* 245 (1964): 245–266; Owen M. Fiss, "Free Speech and Social Structure," *Iowa Law Review* 71 (1985–1986): 1405.

17. Pierre Bourdieu, *The Logic of Practice* (Palo Alto, CA: Stanford University Press, 1990), 131.

18. Margali Sarfatti Larson, *The Rise of Professionalism: A Sociological Analysis* (Berkeley: University of California Press, 1977), 17.

19. Ibid.

20. William J. Goode, "Encroachment, Charlatanism and the Emerging Professions," *American Sociological Review* 25 (1960): 902.

21. Douglas W. Allen, *The Institutional Revolution: Measurement and the Economic Emergence of the Modern World* (Chicago, IL: University of Chicago Press, 2011).

22. James Scott, *Seeing Like a State: How Certain Schemes to Improve the Human Condition Have Failed* (New Haven, CT: Yale University Press, 1999), 2.

23. Arno Mayer, *The Persistence of the Old Regime: Europe to the Great War* (New York: Pantheon Books, 1981).

24. Robert Tavernor, *Smoot's Ear: The Measure of Humanity* (New Haven, CT: Yale University Press, 2007), 49.

25. Ibid., 50.

26. Charles W. Calhoun, ed., *The Gilded Age: Perspectives on the Origins of Modern America* (Lanham, MD: Rowman & Littlefield, 2006), 259.

27. *Interstate Commerce Act (1887), OurDocuments.gov,* http://www .ourdocuments.gov/doc.php?flash=true&doc=49.

28. Hofstadter, *Age of Reform,* 143.

29. Tavernor, *Smoot's Ear,* 143.

30. Scott, *Seeing Like a State,* 3.

31. John Quincy Adams, "Report on Weights and Measures," 1821, http://www.digitalhistory.uh.edu/disp_textbook.cfm?smtID=3 &psid=267.

32. Stephen Kern, *The Culture of Time and Space 1880–1918,* 2nd ed. (Cambridge, MA: Harvard University Press, 2003).

33. Peter Galison, *Einstein's Clocks and Poincaré's Maps: Empires of Time* (New York: W. W. Norton, 2004).

34. Stephen Kern, *The Culture of Time and Space, 1880–1918* (Cambridge, MA: Harvard University Press, 2003), 29.

35. Michael Brian Schiffer, *Power Struggles: Scientific Authority and the Creation of Practical Electricity before Edison* (Cambridge, MA: MIT Press, 2008). See also J. D. Bernal, *Science in History,* vol. 2, *The Scientific and Industrial Revolution* (London: C. A. Watts, 1954; Cambridge, MA: MIT Press, 1985).

36. Thomas L. Haskell, *The Emergence of Professional Social Science: The American Social Science Association and the Nineteenth Century Crisis of Authority* (Champaign: University of Illinois Press, 1977), 26.

37. The term "disassembly line" was coined by Sigfried Giedion; see Daniel J. Boorstin, *The Americans: The Democratic Experience* (New York: Vintage, 1974), chap. 2.

38. Talcott Parsons, "Professions," in *International Encyclopedia of the Social Sciences,* ed. David L. Sills (London: Macmillan, 1968), 536–47.

39. Boorstin, *The Americans.*

40. Oliver Wendell Holmes Sr., "Currents and Counter-Currents in Medical Science," an address delivered before the Massachusetts Medical Society, at the Annual Meeting, May 30, 1860, cited in Burton J. Bledstein, *The Culture of Professionalism: The Middle Class and the Development of Higher Education in America* (New York: W. W. Norton, 1978).

41. Robert Geyer and Samir Rihani, *Complexity and Public Policy: A New Approach to 21st Century Politics, Policy and Society* (Oxford, UK: Routledge, 2010), 20.

42. Brian Cook, *Democracy and Administration: Woodrow Wilson's Ideas and the Challenges of Public Management* (Baltimore, MD: Johns Hopkins University Press, 2007), 129.

43. Frederick Winslow Taylor, *The Principles of Scientific Management* (New York: Harper and Brothers, 1911), 9.

44. Stephen Kern, *Culture of Time and Space,* 316.

45. Andreas Reinstaller and Werner Hölzl, "Big Causes and Small Events: QWERTY and the Mechanization of Office Work," *Industrial and Corporate Change* (May 29, 2009): 999–1031. See also Donald Hoke, "The Woman and the Typewriter: A Case Study in Technological Innovation and Social Change," http://www.thebhc.org/publications /BEHprint/v008/p0076-p0088.pdf.

46. Abbott, *System of Professions,* 1.

47. Thomas L. Haskell, *Objectivity Is Not Neutrality: Explanatory Schemes in History* (Baltimore, MD: Johns Hopkins University Press, 1998), 215.

48. Barbara Ehrenreich, *Fear of Failing: The Inner Life of the Middle Class* (New York: Pantheon Books, 1989), 78.

49. Nicholas R. Parrillo, *Against the Profit Motive: The Salary Revolution in American Government, 1780–1940* (New Haven: Yale University Press, 2013).

50. Renee A. Irvin and John Stansbury, "Citizen Participation in Decision Making: Is It worth the Effort?"*Public Administration Review* 64, no.1 (2004): 58.

51. Michael P. Brown, *Replacing Citizenship: AIDS Activism and Radical Democracy* (New York: Guilford Press, 1997), 137.

52. Cook, *Democracy and Administration,* 131–32.

53. See Congressional Research Service, "History of Civil Service Merit Systems."

54. Eliot Freidson, *Professionalism, the Third Logic: On the Practice of Knowledge* (Chicago, IL: University of Chicago Press, 2001), 22.

55. Boorstin, *The Americans.*

56. Bledstein, *Culture of Professionalism,* 5.

57. Harold Wilensky, cited by Larson in *Rise of Professionalism,* 5.

58. Freidson, *Professionalism,* 22.

59. Edward Purcell Jr., *The Crisis of Democratic Theory: Scientific Naturalism and the Problem of Value* (Lexington, KY: University of Kentucky Press, 1973), 6.

60. William P. LaPiana, *Logic and Experience: The Origin of Modern American Legal Education* (Oxford, UK: Oxford University Press, 1994), 8.

61. Barrington Moore Jr., *Social Origins of Dictatorship and Democracy: Lord and Peasant in the Making of the Modern World* (Boston: Beacon Press, 1966), 112.

62. Bledstein, *Culture of Professionalism,* 1.

63. Hofstadter, *Age of Reform,* 154.

64. Claudia Goldin, "Enrollment in Institutions of Higher Education, by Sex, Enrollment Status, and Type of Institution: 1869–1995," in *Historical Statistics of the United States: Earliest Times to the Present,* vol. 2, part B, Millennial Edition, ed. Susan B. Carter et al. (Cambridge, UK: Cambridge University Press, 2006), Table Bc523-536, 210–13.

65. Purcell, *Originalism,* 7.

66. Morrill Act, 7 U.S.C. § 301 et seq., (1862), 304.

67. William C. Chase, *The American Law School and the Rise of Administrative Government* (Madison: University of Wisconsin Press, 1982), 29.

68. Purcell, *Crisis of Democratic Theory,* 16.

69. Google NGram search, https://books.google.com/ngrams.

70. Crick, *American Science of Politics,* 28.

71. See also Ellen Herman, *The Romance of American Psychology: Political Culture in the Age of Experts* (Berkeley: University of California Press, 1995).

72. Purcell, *Crisis of Democratic Theory,* 25.

73. Ibid., 103.

74. Leonard White, *The Republican Era, 1869–1901: A Study in Administrative History* (New York: Macmillan, 1958), 7–8.

75. Hofstadter, *Age of Reform*, 218.

76. Alexis De Tocqueville, *Democracy in America,* ed. J. P. Mayer, trans. GeorgeLawrence (New York: Doubleday, 1969), 242.

77. Max Weber, "Politics as a Vocation," in *Weber: Political Writings,* ed. Peter Lassman and Ronald Speirs (Cambridge, UK: Cambridge University Press, 1994), 345.

78. Ari Hoogenboom, *Outlawing the Spoils: A History of the Civil Service Reform Movement, 1865–1883* (Urbana: University of Illinois Press, 1961). See also Congressional Research Service, "History of Civil Service Merit Systems."

79. Leonard White, *The Jeffersonians: A Study in Administrative History, 1801–1829* (New York: Macmillan, 1951), 368.

80. Parrillo, *Against the Profit Motive.*

81. Robert Michels, *Political Parties: A Sociological Study of the Oligarchical Tendencies of Modern Democracy* (New York: Hearst International Library, 1915), 186, available as a Google eBook at http://bit.ly/1vxbTK7.

82. Daniel P. Carpenter, *The Forging of Bureaucratic Autonomy: Reputations, Networks, and Policy Innovation in Executive Agencies, 1862–1928* (Princeton, NJ: Princeton University Press, 2001), 41.

83. White, *Republican Era*, 18.

84. Mashaw, *Creating the Administrative Constitution*, 21.

85. Theodore J. Lowi, *The End of Liberalism* (New York: W. W. Norton, 1969), 128–29.

86. Woodrow Wilson, *The Study of Public Administration* (Washington, DC: Public Affairs Press, 1955).

87. Mashaw, *Creating the Administrative Constitution*, 238–39.

88. White, *Republican Era*, 387.

89. Bledstein, *Culture of Professionalism,* 122.

90. White, *Republican Era*, 396.

91. Woodrow Wilson, "Of the Study of Politics," *New Princeton Review* 3 (1887): 188–99.

92. James Q. Wilson, *Bureaucracy: What Government Agencies Do and Why They Do It* (New York: Basic Books, 1989), 377.

93. Nicholas Christakis, "Let's Shake Up the Social Sciences," *New York Times,* July 19, 2013.

94. Fritz Ringer, *The Decline of the German Mandarins: The German Academic Community 1890–1938* (Middletown, CT: Wesleyan University Press, 1990) (chronicling the rise of the educated, middle-class elite in Germany known as the mandarins).

95. Harold Laski, "The Limitations of the Expert," in *Man and the State: Modern Political Ideas,* ed. William Ebenstein (New York: Rinehart, 1947), 148; Reprinted from Harold Laski, "The Limitations of the Expert," *Harper's Monthly,* December 1930, 101.

96. Scott, *Seeing Like a State,* 21.

97. Carpenter, *Forging of Bureaucratic Autonomy,* 3.

98. Haskell, *Emergence of Professional Social Science,* 26.

3. The Limits of Democratic Theory

Epigraph: Walter Lippmann, *The Phantom Public* (New York: Harcourt, Brace, 1925), 13–14.

1. "We the People, Your Voice in Government," Petitions, The White House, https://petitions.whitehouse.gov/.

2. John W. Kingdon, *Agendas, Alternatives, and Public Policies,* Longman Classics ed. (London: Longman, 2002), 3.

3. Leigh Heyman, "There's Now an API for We the People" (blog), *The White House,* May 1, 2013, http://www.whitehouse.gov/blog/2013/05/01/theres-now-api-we-people.

4. Dave Karpf, "How the White House's E-People Petition Site Became a Virtual Ghost Town," *TechPresident,* June 20, 2014, http://techpresident.com/news/25144/how-white-houses-we-people-e-petition-site-became-virtual-ghost-town.

5. Audie Cornish, "White House's 'We the People' Petitions Find Mixed Success," National Public Radio, January 3, 2013, http://www.npr.org/2013/01/03/168564135/white-houses-we-the-people-petitions-find-mixed-success; Heyman, "There's Now an API"; Tom McKay, "Death Star Petition Another Sign of Broken White House 'We the People' System," *PolicyMic,* December 28, 2012, at: http://www.policymic.com/articles/21574/death-star-white-house-petition-another-sign-of

-broken-we-the-people-system; Macon Phillips, "Why We're Raising the Signature Threshold for We the People" (blog), *The White House*, January 15, 2013, http://www.whitehouse.gov/blog/2013/01/15/why-we-re-raising-signature-threshold-we-people; Macon Phillips, "Sunshine Week: In Celebration of Civic Engagement" (blog), *The White House*, March 13, 2013, http://www.whitehouse.gov/blog/2013/03/13/sunshine-week-celebration-civic-engagement; Peter Welsch, "Civic Hacking at the White House: We the People, by the People" (blog), *The White House*, June 5, 2013, http://www.whitehouse.gov/blog/2013/06/05/civic-hacking-white-house-we-people-people.

6. James Madison, "Federalist 57," *The Federalist Papers* (Toronto: Bantam Books, 1982). "The aim of every political constitution is, or ought to be, first to obtain for rulers men who possess most wisdom to discern, and most virtue to pursue, the common good of the society; and in the next place, to take the most effectual precautions for keeping them virtuous whilst they continue to hold their public trust."

7. Harold Laski, "The Limitations of the Expert," in *Man and the State: Modern Political Ideas,* ed. William Ebenstein (New York: Rinehart, 1947), 159ff.

8. E. Donald Elliott, "Re-Inventing Rulemaking," *Duke Law Journal* 41 (1992): 1490, 1492.

9. Frank Fischer, *Democracy and Expertise: Reorienting Policy Inquiry* (Princeton, NJ: Princeton University Press, 2009), Kindle ed,, location 695. Fischer cites especially D. Zolo, *Complexity and Democracy: A Realist Approach* (Cambridge, UK: Polity Press, 1992); Claus Offe, *Modernity and the State: East, West* (Cambridge, UK: Polity Press, 1996); and Jürgen Habermas, *Between Facts and Norms: Contributions to a Discourse Theory of Law and Democracy,* trans. W. Rehg (Cambridge, MA: MIT Press).

10. Robert A. Dahl, *Democracy and Its Critics* (New Haven, CT: Yale University Press, 1989), 10.

11. Cary Coglianese, "Citizen Participation Rulemaking: Past, Present, and Future," *Duke Law Journal* 55 (2006): 951.

12. Hollie Russon-Gilman, "The Participatory Turn: Participatory Budgeting Comes to the United States" (PhD diss., Government

Department, Harvard University, 2012), 24. See also Hollie Russon
Gilman, "Participatory Budgeting Year in Review: Looking Back to
the Future," *The GovLab* (blog), December 23, 2013, http://thegovlab
.org/participatory-budgeting-year-in-review-looking-back-to-the
-future/; Hollie Russon Gilman, "New York's Big Participatory
Budgeting Moment," *The GovLab* (blog), November 3, 2013, http://
thegovlab.org/new-yorks-big-participatory-budgeting-moment/;
Hollie Russon Gilman, "Can Citizens Make Collaborative Public
Decisions?" GovLab, June 17, 2014; Hollie Russon Gilman, "Boston
Teens Take on the Budget," *Boston Globe,* July 16, 2014.

13. Archon Fung, *Empowered Participation: Reinventing Urban Democ-
racy* (Princeton, NJ: Princeton University Press, 2006). See also
Archon Fung and Erik Olin Wright, eds., *Deepening Democracy:
Institutional Innovations in Empowered Participatory Governance*
(Brooklyn, NY: Verso, 2003).

14. Evgeny Morozov, "The Internet Ideology: Why We Are allowed to
Hate Silicon Valley," FAZ.NET, November 11, 2013, sec. Feuilleton,
http://www.faz.net/aktuell/feuilleton/debatten/the-internet-ideology
-why-we-are-allowed-to-hate-silicon-valley-12658406.html.

15. Geoff Mulgan, "Ten Predictions for 2015" (blog), *NESTA,* December
2014, http://www.nesta.org.uk/news/2015-predictions/democracy
-makes-itself-home-online.

16. Joseph E. Stiglitz, "In No One We Trust" (blog), *Opinionator, New
York Times,* December 21, 2013, http://opinionator.blogs.nytimes.com
/2013/12/21/in-no-one-we-trust/.

17. Stephen P. Turner, *Liberal Democracy 3.0: Civil Society in an Age of
Experts* (London: Sage, 2003), 12.

18. Ilya Somin, *Democracy and Political Ignorance: Why Smaller Govern-
ment Is Smarter* (Stanford, CA: Stanford University Press, 2013); Rick
Shenkman, *Just How Stupid Are We? Facing the Truth about the
American Voter* (New York: Basic Books, 2008).

19. David Estlund, "Why Not Epistocracy?" in *Desire, Identity and
Existence: Essays in Honor of T. M. Penner*, ed. Naomi Reshotko
(Kelowna, BC: Academic, 2003).

20. Leonard D. White, *The Federalists: A Study in Administrative History*
(New York: Macmillan, 1961).

21. G. Pincione and F. Tesón, *Rational Choice and Democratic Deliberation: A Theory of Discourse Failure* (Cambridge, UK: Cambridge University Press, 2006).

22. Bryan Caplan, *The Myth of the Rational Voter: Why Democracies Choose Bad Policies* (Princeton, NJ: Princeton University Press, 2007).

23. Somin, *Democracy and Political Ignorance,* 3–4.

24. Carole Pateman, *Participation and Democratic Theory* (Cambridge, UK: Cambridge University Press, 1970), 109.

25. Somin, *Democracy and Political Ignorance,* 41.

26. Hiroki Azuma, *General Will 2.0: Rousseau, Freud, Google* [Ippan Ishi 2.0—Ruso, Furoito, Guguru], trans. John Person and Naomi Matsuyama (New York: Vertical, 2014), Kindle ed., location 1049.

27. Seymour Martin Lipset, *Political Man: The Social Bases of Politics* (Baltimore, MD: Johns Hopkins University Press, 1981), 45.

28. Harry C. Boyte, "Reframing Democracy: Governance, Civic Agency, and Politics," *Public Administration Review* 65, no. 5 (2005): 536–46.

29. Dorwin Cartwright and Alvin Zander, *Group Dynamics: Research and Theory,* 3rd ed. (New York: Harper, 1968; Jonathon Gillette and Marion McCollom, eds., *Groups in Context: A New Perspective on Group Dynamics* (Reading, MA: Addison-Wesley, 1990); Cass R. Sunstein, *Why Societies Need Dissent* (Cambridge, MA: Harvard University Press, 2003).

30. Cass R. Sunstein and Reid Hastie, "Garbage In, Garbage Out? Some Micro Sources of Macro Errors," *Journal of Institutional Economics* 10 (March 2014): 1–23; Cass R. Sunstein, "Deliberative Trouble? Why Groups Go to Extremes," *Yale Law Journal* 110, no. 71 (2000): 1–200; Cass R. Sunstein, "The Law of Group Polarization," in *Debating Deliberative Democracy,* ed. James Fishkin and Peter Laslett (Malden, MA: Blackwell, 2003), 80–102; See also Richard Posner, *Law Pragmatism and Democracy* (Cambridge, MA: Harvard University Press, 2003); and Dan Hunter, "Philippic.com," *California Law Review* 90, no. 611 (2002): 640 ("If there is any news more startling than that a Macedonian was triumphing over Athens, then it is surely that a leading democratic theorist should argue that group discussion and deliberation are bad things. Startling though this may be, it is nonetheless true. The central fear of REPUBLIC.COM is a fear of

groups. More particularly, Sunstein fears the polarizing effect of groups upon the decision-making and thinking of its members.").

31. Lawrence J. Sanna and Craig D. Parks, "Group Research Trends in Social and Organizational Psychology: Whatever Happened to Intragroup Research?" *Psychological Science* 8 (1997): 261–67; D. Stokols et al., "The Science of Team Science: Overview of the Field and Introduction to the Supplement," *American Journal of Preventive Medicine* 35 (2008): S77–S89.

32. Kurt Lewin, "Research Center for Group Dynamics at Massachusetts Institute of Technology," *Sociometry* 8, no. 2 (1945): 126–36. The center became an intellectual incubator for the design of imaginative studies of group communication; see, e.g., A. Bavelas, "Communication Patterns in Task-Oriented Groups," *Journal of Acoustical Society of America* 22 (1950): 725–30; Harold Guetzkow and John Gyr, "An Analysis of Conflict in Decision-making Groups," *Human Relations* 7 (1954): 367–81; Harold Guetzkow and Herbert A. Simon, "Mechanisms Involved in Group Pressures on Deviate Members," *British Journal of Statistical Psychology* 8, (1955): 93–102; Harold J. Leavitt, "Some Effects of Certain Communication Patterns on Group Performance," *Journal of Abnormal and Social Psychology* 46, no. 1 (1951): 38–50; Harold J. Leavitt and Ronald A. H. Mueller, "Some Effects of Feedback on Communication," *Human Relations* 4 (November 1951): 401–10, which generated hundreds of published articles. See also Peter R. Monge and Soshir S. Contractor, *Theories of Communications Networks* (Oxford, UK: Oxford University Press, 2003).

33. J. Richard Hackman and Nancy Katz, "Group Behavior and Performance" in *The Handbook of Social Psychology,* ed. Susan T. Fiske, Daniel T. Gilbert, and Gardner Lindzey (New York: John Wiley and Sons, 2010), 1220; J. R. Hackman, *Groups That Work (and Those That Don't): Creating Conditions for Effective Teamwork* (San Francisco: Jossey-Bass, 1990).

34. National Research Council, *Using Science as Evidence in Public Policy* (Washington, DC: National Academies Press, 2012), http://www.nap.edu/openbook.php?record_id=13460&page=53. See also A. Pentland, "The New Science of Building Great Teams," *Harvard Business Review* 90 (2012): 60–69.

35. Daniel Kahneman, *Thinking, Fast and Slow* (New York: Farrar, Straus and Giroux, 2013); D. Kahneman and G. Klein, "Conditions for Intuitive Expertise: A Failure to Disagree," *American Psychologist* 64, no. 6 (2009): 515–26; D. Kahnemann, P. Slonic, and A. Tversky, eds., *Judgment under Uncertainty: Heuristics and Biases* (New York: Cambridge University Press, 1982).

36. National Research Council. *Using Science as Evidence in Public Policy.*

37. Vincent Price, Lilach Nir, Joseph N. Capella, "Normative and Informational Influences in Online Political Discussions," *Communication Theory* 16, no. 1 (2006): 47–74 (studying the data from sixty online group discussions, involving ordinary citizens, about the tax plans offered by rival U.S. presidential candidates George W. Bush and Al Gore in 2000).

38. M. D. Conover et al., "Political Polarization on Twitter," in *Proceedings of the Fifth International AAAI Conference on Weblogs and Social Media* (Menlo Park, CA: AAAI Press, 2011); A. Lada and Natalie Glance, "The Political Blogosphere and the 2004 U.S. Election: Divided They Blog," in *Proceedings of the Third International Workshop on Link Discovery* (New York: ACM, 2005), 36–43; Cass R. Sunstein, "The Law of Group Polarization," *Journal of Political Philosophy* 10, no. 2 (2002):175–95; Cass R. Sunstein, *Republic.com 2.0* (Princeton, NJ: Princeton University Press, 2007); Sarita Yardi and Danah Boyd, "Dynamic Debates: An Analysis of Group Polarization over Time on Twitter," *Bulletin of Science, Technology and Society* 20 (2010): S1–S8.

39. Itai Himelboim, "Mapping Twitter Topic Networks: From Polarized Crowds to Community Clusters," Pew Research Center's Internet & American Life Project, February 20, 2014, http://www.pewinternet .org/2014/02/20/mapping-twitter-topic-networks-from-polarized -crowds-to-community-clusters/.

40. Price, Nir, and Capella, "Normative and Informational Influences"; Neil Weinstock Netanel, "Cyberspace Self-Governance: A Skeptical View from Liberal Democratic Theory," *California Law Review* 88 (2000): 395.

41. Michelle Andrews et al., "Mobile Ad Effectiveness: Hyper-Contextual Targeting with Crowdedness" (Fox School of Business Research Paper No. 15-040, 2015), 1–17, http://dx.doi.org/10.2139/ssrn.2439388.

42. Beth Simone Noveck, "A Democracy of Groups," *First Monday* 10, no. 11 (2005), http://firstmonday.org/article/view/1289/1209.

43. On the affinities between the New Deal regulation of the economy and the principles of mass production, see Michael J. Piore and Charles F. Sabel, *The Second Industrial Divide: Possibilities for Prosperity* (New York: Basic Books, 1984): 73–104.

44. Adrian Bejan, *Design in Nature: How the Constructal Law Governs Evolution in Biology, Physics, Technology, and Social Organization* (New York: Knopf Doubleday, 2012).

45. Aristotle, *Politics,* Book 7, sec. 1326a, http://www.perseus.tufts.edu/hopper/text?doc=urn:cts:greekLit:tlg0086.tlg035.perseus-eng1:7.1326a.

46. John Stuart Mill, *Considerations on Representative Government* (London: Longman, 1865).

47. Oliver E. Williamson, "The Organization of Work: A Comparative Institutional Assessment," *Journal of Economic Behavior and Organization* 1 (1980): 35. See also Ronald H. Coase, "The Nature of the Firm" (1937), reprinted in Oliver E. Williamson and Sidney G. Winter, *The Nature of the Firm: Origins, Evolution and Development* (New York: Oxford University Press, 1991), 18–33. See also Oliver E. Williamson, *Markets and Hierarchies: Analysis and Antitrust Implications* (New York: Free Press, 1983); Oliver E. Williamson, "Transaction-Cost Economics: The Governance of Contractual Relations," *Journal of Law and Economics* 22, nos. 233, 234 (1979): 250–53.

48. Robert Michels, *Political Parties: A Sociological Study of the Oligarchical Tendencies of Modern Democracy* (New York: Hearst International Library, 1915), 191.

49. See Richard Hofstadter, *The Age of Reform* (New York: Vintage Books, 1955), 217.

50. Michels, *Political Parties,* 65.

51. Brian J. Cook, *Democracy and Administration: Woodrow Wilson's Ideas and the Challenges of Public Management* (Baltimore, MD: Johns Hopkins University Press, 2007), 131–32.

52. Robert K. Merton, *Social Theory and Social Structure,* rev. ed. (New York: Free Press, 1957).

53. Nassim Nicholas Taleb, *Antifragile: Things that Gain from Disorder* (New York: Random House, 2012).

54. Over time, this shift in focus toward representative democracy gave rise to government bureaucracy in America, as the number and diversity of groups and interests that needed representation dramatically increased, with the consequence of physically separating the people from their government and making the act of governing a distant and closed process for most citizens. See Robert A. Dahl, *Democracy and Its Critics* (New Haven, CT: Yale University Press, 1989).

55. For an extensive catalog of crowdsourcing examples, see "The GovLab Compendium," *GovLab Research* (Spring 2013), http://www.thegovlab .org/wp-content/uploads/2013/04/GovLabCompendium.pdf.

56. Thomas Christiano, "The Authority of Democracy," *Journal of Political Philosophy* 12 (2004): 266–90.

57. Simone Chambers, "Deliberative Democratic Theory," *Annual Review of Political Science* 6, no. 1 (2003): 307–26.

58. Habermas, *Between Facts and Norms.*

59. James S. Fishkin, *The Voice of the People: Public Opinion and Democracy* (New Haven, CT: Yale University Press, 1997); James Fishkin, *When the People Speak: Deliberative Democracy and Public Consultation* (Oxford, UK: Oxford University Press, 2009).

60. James S. Fishkin and Robert C. Luskin, "Experimenting with a Democratic Ideal: Deliberative Polling and Public Opinion," *Acta Politica* 40, no. 3 (2005): 284–98.

61. See, e.g., James Fishkin and Robert Luskin, "The Quest for Deliberative Democracy," *The Good Society* 9(1) (1999): 1–9; Amy Gutmann and Dennis Thompson, *Why Deliberate Democracy?* (Princeton, NJ: Princeton University Press, 2004); Joshua Cohen, "Deliberation and Democratic Legitimacy," in *The Good Polity*, ed. Alan Hamlin and Philip Pettit (Oxford, UK: Oxford University Press, 1989), 17–35.

62. G. Rowe and L. J. Frewer "Public Participation Methods: A Framework for Evaluation," *Science, Technology and Human Values* 25, no. 1 (2000): 3–29.

63. Elizabeth Anderson, "An Epistemic Defense of Democracy: David Estlund's Democratic Authority," *Episteme* 5, no. 1 (2008): 129–39; Elizabeth Anderson, "The Epistemology of Democracy," *Episteme* 3, nos. 1–2 (2006): 8–22; David Estlund, *Democratic Authority: A Philosophical Framework* (Princeton, NJ: Princeton University Press,

2009); Estlund, "Why Not Epistocracy?"; Helene Landemore, *Democratic Reason: Politics, Collective Intelligence, and the Rule of the Many* (Princeton, NJ: Princeton University Press, 1999).

64. For a good summary of epistemic democracy, see also Cathrine Holst and Anders Molander, "Epistemic Democracy and the Accountability of Experts," in *Expertise and Democracy,* ed. Cathrine Holst (ARENA Report 1/14, Arena Centre for European Studies, University of Oslo, Norway, February 2014).

65. Marquis de Condorcet, "On Elections," in *Condorcet: Foundations of Social Choice and Political Theory,* trans. and ed. Iain McLean and Fiona Hewitt (Brookfield, VT: Edward Elgar, 1994).

66. Geoff Mulgan, "True Collective Intelligence: A Sketch of a Possible New Field," *Philosophy and Technology* 27 (2014): 133–42.

4. The Technologies of Expertise

Epigraph: T. S. Eliot, *The Rock: A Pageant Play Written for Performance on Behalf of the Forty-Five Churches Fund of the Diocese of London* (London: Faber and Faber, 1934).

1. John Murdoch-Burn, "Study: Less than 1% of the World's Data Is Analysed, Over 80% Is Unprotected," *The Guardian,* December 19, 2012, http://www.theguardian.com/news/datablog/2012/dec/19/big -data-study-digital-universe-global-volume.

2. McKinsey Global Institute, *Big Data: The Next Frontier for Innovation, Competition, and Productivity,* Technical Report, June 2011. See also M.G. Siegler, "Eric Schmidt: Every 2 Days We Create as Much Information as We Did Up to 2003," *Tech Crunch,* August 4, 2010, http://techcrunch.com/2010/08/04/schmidt-data/.

3. Clive Thompson, *Smarter Than You Think: How Technology Is Changing Our Minds for the Better* (New York: Penguin, 2013), 47.

4. See Jacques Bughin, Michael Chui, and James Manyika, "Clouds, Big Data, and Smart Assets: Ten Tech-Enabled Business Trends to Watch," McKinsey Global Institute, September 22, 2010, http://www .mckinsey.com/insights/mgi/in_the_news/clouds_big_data_and _smart_assets.

5. Antoinette Schoar and Dutta Sagatto, "The Power of Heuristics" (working paper, Ideas42, January 2014), 1–8, http://www.ideas42.org /publication/view/the-power-of-heuristics/.

6. Peter D. Blair, "Scientific Advice for Policy in the United States," in *The Politics of Scientific Advice: Institutional Design for Quality Assurance,* ed. Justus Lentsch and Peter Weingart (Cambridge, UK: Cambridge University Press, 2011), 298.

7. See J. M. Balkin, *Cultural Software: A Theory of Ideology* (New Haven, CT: Yale University Press, 2003).

8. David Weinberger, *Too Big to Know: Rethinking Knowledge Now That the Facts Aren't the Facts, Experts Are Everywhere, and the Smartest Person in the Room Is the Room* (New York: Basic Books, 2012), vii.

9. P. M. Wilson et al., "The Expert Patients Programme: A Paradox of Patient Empowerment and Medical Dominance," *Health and Social Care in the Community* 15, no. 5 (2007): 426–38.

10. Alexandra Horowitz, *On Looking: Eleven Walks through Expert Eyes* (New York: Scribner, 2013).

11. John Dewey, *The Public and Its Problems* (New York: H. Holt and Sons, 1927; Pennsylvania State University Press, 2012), 20.

12. Bo-Christer Björk, Annikki Roos, and Mari Lauri, "Scientific Journal Publishing: Yearly Volume and Open Access Availability," *Information Research* 14, no. 1 (2009), http://informationr.net/ir/14–1/paper391 .html.

13. Emily Rydel Ackman, *Getting Scholarship into Policy: Lessons from University-Based Bipartisan Scholarship Brokers* (PhD diss., University of Arizona, 2013). See also, A. F. Ball, "To Know Is Not Enough: Knowledge, Power, and the Zone of Generativity," *Educational Researcher* 41, no. 8 (2012): 283–93.

14. See, e.g., "Executive Suite Interview with the Collaborative and Footnote," February 16, 2014, http://collaborativeri.org/index.php /executive-suite-interview-with-the-collaborative-and-footnote/.

15. Michael Staton, "The Degree Is Doomed," *Harvard Business Review* Blog Network, January 8, 2014, http://blogs.hbr.org/2014/01/the -degree-is-doomed/.

16. Frank Fisher, *Democracy and Expertise: Reorienting Policy Inquiry* (Princeton, NJ: Princeton University Press, 2009), Kindle ed., location 56.

17. Dorothy Leonard and Walter Swap, "Deep Smarts: Experience-Based Wisdom," Center for Association Leadership, Executive Update, January 2005, http://www.asaecenter.org/Resources/EUArticle.cfm ?ItemNumber=11523.

18. Dmitry Gubanov et al., *E-Expertise: Modern Collective Intelligence* 558, Studies in Computational Intelligence (Heidelberg, Germany: Springer, 2014), xvii.

19. H. Dreyfus and S. Dreyfus, "Expertise in Real World Contexts," *Organization Studies* 26, no. 5 (2005): 779–92; Anders K. Ericsson et al., ed., *Cambridge Handbook on Expertise and Expert Performance* (Cambridge, UK: Cambridge University Press, 2006); Anders K. Ericsson, Michael J. Prietula, and Edward T. Cokely, "The Making of an Expert," *Harvard Business Review* (July–August 2007), https://hbr .org/2007/07/the-making-of-an-expert; Harald A. Mieg, *The Social Psychology of Expertise* (Mahwah, NJ: Lawrence Erlbaum, 2001); J. Shanteau et al., "Performance-Based Assessment of Expertise: How to Decide if Someone Is an Expert or Not," *European Journal of Operational Research* 136, no. 2 (2001): 253–63.

20. Jon Taplin, "Talent Matters," July 15, 2014, http://jontaplin.com/2014 /07/15/talent-matters/. Brooke N. Macnamara, David Z. Hambrick, and Frederick L. Oswald, "Deliberate Practice and Performance in Music, Games, Sports, Education, and Professions: A Meta-Analysis," *Psychological Science* (August 1, 2014): 1608–18.

21. Harry Collins, *Are We All Scientific Experts Now?* (London: Polity Press, 2014).

22. Harry Collins and Charles Evans, *Rethinking Expertise* (Chicago, IL: University of Chicago Press, 2007). Collins and Evans refined their table in subsequent work.

23. Krisztian Balog, et al., "Expertise Retrieval," in *Foundations and Trends in Information Retrieval* 6, nos. 2–3 (2012), 127–256c.

24. HeyPress, https://www.hey.press/; Gradberry, http://techcrunch.com /2015/03/03/from-pakistan-to-y-combinator-gradberry-vets-technical -talent/; CyberCompEx, https://www.cybercompex.org/.

25. These so-called World Wide Web Consortium (W3C) and other experiments are described in Balog et al., "Expertise Retrieval."

26. John Corson-Rikert, "Map of VIVO Projects around the World," *Duraspace,* July 28, 2013, https://wiki.duraspace.org/display/VIVO

/VIVOmap. See also Mike Conlon, "Scholarly Networking: Needs and Desires," in *VIVO: A Semantic Approach to Scholarly Networking and Discovery*, ed. Katy Boerner et al. (San Rafael, CA: Morgan and Claypool, 2012).

27. Indiana Database of University Research Expertise (INDURE), www.indure.org.

28. James Powell et al., "EgoSystem: Where Are Our Alumni?" *Code4Lib*, no. 24 (April 2014), http://journal.code4lib.org/articles/9519.

29. Sao Paulo Research Foundation, "Research in Sao Paulo," FAPESP, http://www.fapesp.br/en/5385. Sao Paulo Research Foundation, "Instructions for Elaborating the Summary of the CV of the Principal Investigator," FAPESP, April 26, 2012, http://www.fapesp.br/en/6351.

30. M. Wedel and W. Kamakura, *Market Segmentation: Conceptual and Methodological Foundations* (Norwell, MA: Kluwer Academic Publishers, 2000).

31. "Big Data and Advanced Analytics: 16 Use Cases from a Practitioner's Perspective," Workshop at a NASSCOM Conference, June 27, 2013, http://www.slideshare.net/McK_CMSOForum/big-data-and-advanced-analytics-16-use-cases.

32. Steinbrink, Susan, "Behavioral Targeting: The Future of Online Advertising," PhoCusWright Innovation, September 2012, 1–11, http://www.phocuswright.com/Travel-Research/Technology-Innovation/Behavioral-Targeting-The-Future-of-Online-Advertising#.VSm2CSdlyuQ.

33. Dagar et al., "Internet, Economy and Privacy," https://www.funginstitute.berkeley.edu/sites/default/files/Internet-Economy-and-Privacy.pdf.

34. Hamilton Consultants, "Economic Value of the Advertising-Supported Internet Ecosystem," June 10, 2009, http://www.iab.net/media/file/Economic-Value-Report.pdf.

35. Marc-André Thibodeau, Simon Bélanger, Claude Frasson, "WHITE RABBIT—Matchmaking of User Profiles Based on Discussion Analysis Using Intelligent Agents," *Intelligent Tutoring Systems Lecture Notes in Computer Science* 1839 (2000): 113–22.

36. See, for example, Jennifer Valentino-Devries and Jeremy Singer-Vine, "They Know What You're Shopping For," *Wall Street Journal*, December 7, 2012, http://online.wsj.com/article/SB10001424127887324

78440457814314413273 6214.html. "Dataium LLC, the company that can track car shoppers . . . Dataium said that shoppers' Web browsing is still anonymous, even though it can be tied to their names. The reason: Dataium does not give dealers click-by-click details of people's Web surfing history but rather an analysis of their interests. A *Wall Street Journal* examination of nearly 1,000 top websites found that 75 percent now include code from social networks, such as Facebook's 'Like' or Twitter's 'Tweet' buttons. Such code can match people's identities with their Web-browsing activities on an unprecedented scale and can even track a user's arrival on a page if the button is never clicked." Gresi Sanje and Isil Senol, "The Importance of Online Behavioral Advertising for Online Retailers," *International Journal of Business and Social Science* 3, no. 18 (September 2012): 114–21; Derrick Harris, "You Might Also Like . . . to Know-how Online Recommendations Work," *GigaOM,* January 29, 2013, http://gigaom.com/2013/01/29 /you-might-also-like-to-know-how-online-recommendations-work/. Predicting user behavior through modeling based on interests and activity in comparison to similar users is a widespread practice— Netflix, Amazon, Eventbrite. "Netflix claims that 75 percent of what people watch comes from some sort of a recommendation."

37. Mike Hess and Pete Doe, "The Marketer's Dilemma: Focusing on a Target or a Demographic? The Utility of Data-Integration Techniques," *Journal of Advertising Research* 53, no. 2 (June 2013): 231–36.

38. Charles Duhigg, "How Companies Learn Your Secrets," *New York Times Magazine,* February 16, 2012, http://www.nytimes.com/2012/02 /19/magazine/shopping-habits.html.

39. See Tianyi Jiang and Alexander Tuzhilin, "Improving Personalization Solutions through Optimal Segmentation of Customer Bases," *IEEE Transactions on Knowledge and Data Engineering* 21, no. 3 (March 2009) (detailing the shift from statistical to transactional data mining for customer segmentation). Cosimo Palmisano, Alexander Tuzhilin, and Michele Gorgoglione, "Using Context to Improve Predictive Modeling of Customers in Personalization Applications," *IEEE Transactions on Knowledge and Data Engineering* 20, no. 11 (November 2008) (concluding that context matters for predicting individual behavior of customers but not groups of customers in the aggregate); T. Jiang and A. Tuzhilin, "Segmenting Customers from

Population to Individuals: Does 1 to 1 Keep Your Customers Forever?" *IEEE Transactions on Knowledge and Data Engineering* 18, no. 10 (October 2006), 1297–1311.

40. Eli Pariser, *The Filter Bubble: What the Internet Is Hiding from You* (New York: Penguin Press HC, 2011); D. J. Patil, *Data Jujitsu: The Art of Turning Data into Product* (Sebastopol, CA: O'Reilly Media, 2012).

41. Shannon David, "How to Deepen Your Impact by Engaging Skilled Volunteers," *Engaging Volunteers,* July 7, 2014, http://blogs .volunteermatch.org/engagingvolunteers/2014/07/07/how-to-deepen -your-impact-by-engaging-skilled-volunteers/.

42. Al M. Rashid et al., "Motivating Participation by Displaying the Value of Contribution," *Proceedings of the SIGCHI Conference on Human Factors in Computing Systems* (Montreal: CHI, 2006), 955.

43. Beth Simone Noveck, *Wiki Government: How Technology Can Make Government Better, Democracy Stronger, and Citizens More Powerful* (Washington, DC: Brookings Institution Press, 2009), 80ff.

44. Joan Morris DiMicco, "Changing Small Group Interaction through Visual Reflections of Social Behavior" (Ph.D. diss., MIT, 2005); Joan DiMicco and Walter Bender, "Group Reactions to Visual Feedback Tools" (paper prepared for the Second International Conference on Persuasive Technology, April 26–27, 2007, Stanford University); DiMicco's papers are available at www.joandimicco.com.

45. LinkedIn, "Recommendations Overview: How Do Recommendations Work?" May 4, 2014, http://help.linkedin.com/app/answers/detail/a _id/90.

46. Patil, *Data Jujitsu*.

47. A. Pal, S. Chang, and J. Konstan, "Evolution of Experts in Question Answering Communities," in *Proceedings of the International AAAI Conference on Weblogs and Social Media* (Association for the Advancement of Artificial Intelligence, 2012), 274–81, https://www.aaai .org/ocs/index.php/ICWSM/ICWSM12/paper/viewFile/4653/4993. See also F. Riahi et al., "Finding Expert Users in Community Question Answering," in *Proceedings of the International Conference on World Wide Web—Workshop on Community Question Answering on the Web* (CQA'12) (New York: ACM, 2012), 791–798.

48. G. Ipeirotis Pagagiotis and Evgeniy Gabrilovich, "Quizz: Targeted Crowdsourcing with a Billion (Potential) Users," in *Proceedings of the*

23rd International Conference on World Wide Web (WWW '14) (Geneva, Switzerland: International World Wide Web Conferences Steering Committee, 2014), 143–54.

49. Balog et al., "Expertise Retrieval," 236.

50. Jamie Heywood, "The Big Idea My Brother Inspired," *TedTalks,* 2009, http://www.ted.com/talks/jamie_heywood_the_big_idea_my _brother_inspired .

51. "About VIVO," VIVO website, http://www.vivoweb.org/about.

52. Anurag Acharya, interview with the author, July 29, 2013, Mountain View, CA.

53. K. Ehrlich, C. Lin, and V. Griffiths-Fisher, "Searching for Experts in the Enterprise: Combining Text and Social Network Analysis," in *Proceedings of the 2007 International ACM SIGGROUP Conference on Supporting Group Work* (GROUP '07) (New York: ACM, 2007), 117–126.

54. Charles D Borromeo, Titus K. Schleyer, Michael J. Becich, and Harry Hochheiser, "Finding Collaborators: Toward Interactive Discovery Tools for Research Network Systems," *Journal of Medical Internet Research* 16, no. 11 (2014), http://www.jmir.org/2014/11/e244/.

55. Relationship Science, "The Relationship Capital Platform," https://www.relsci.com/.

56. Jack Klout, "Don't Count Klout Out: 4 Reasons It's the Platform to Watch in 2014," *Adotas,* March 26, 2014, http://www.adotas.com /2014/03/don%E2%80%99t-count-klout-out-4-reasons-it%E2%80%99s -the-platform-to-watch-in-2014/. See also *The Klout Blog,* http://blog .klout.com.

57. Mark Epernay (aka John Kenneth Galbraith), *The MacLandress Solution* (New York: Houghton Mifflin, 1963), 2. See also Gore Vidal, "Citizen Ken," *New York Review of Books,* December 12, 1963, http://www.nybooks.com/articles/archives/1963/dec/12/citizen-ken/.

58. Mariano Beguerisse-Díaz et al., "Interest Communities and Flow Roles in Directed Networks: The Twitter Network of the UK Riots,"*Journal of the Royal Society Interface* 11 (October 8, 2014), DOI: 10.1098/rsif.2014.0940.

59. Once credentials are granted, they are rarely revoked. This means that credentials must be evaluated for timeliness in order to provide the best results. See J. Shanteau et al., "How Can You Tell Someone Is an

Expert? Empirical Assessment of Expertise," in *Emerging Perspectives on Judgment and Decision Research,* ed. L. Sandra, S. L. Schneider, and J. Shanteau (Cambridge, UK: Cambridge University Press, 2003).

60. Cass R. Sunstein, "The Storrs Lectures: Behavioral Economics and Paternalism," *Yale Law Journal* 122, no. 7 (May 2013), http://www .yalelawjournal.org/feature/the-storrs-lectures-behavioral-economics -and-paternalism. See also, e.g., Ian Ayres and Robert Gertner, "Filling Gaps in Incomplete Contracts: An Economic Theory of Default Rules," *Yale Law Journal* 99, no. 87 (1989): 91 (coining the term "penalty default" to refer to contract default rules that are designed to force parties to bargain over a specific subject); Charles J. Goetz and Robert E. Scott, "The Mitigation Principle: Toward a General Theory of Contractual Obligation," *Virginia Law Review* 69 (1983): 967 (discussing information-forcing default rules).

61. Cory Ondrejka, "Escaping the Gilded Cage: User Created Content and Building the Metaverse," in *The State of Play: Law and Virtual Worlds,* ed. Jack M. Balkin and Beth Simone Noveck (New York: NYU Press, 2006) (the average user in Second Life spends more than an hour a week modifying his or her avatar).

62. "The Upshot: Here's What Will Truly Change Higher Education: Online Degrees That Are Seen as Official," *New York Times,* March 5, 2015, http://www.nytimes.com/2015/03/08/upshot/true-reform-in -higher-education-when-online-degrees-are-seen-as-official.html?ref =opinion&_r=1&abt=0002&abg=1. For example, Basno is one site that allows people to create and collect authenticated badges; see http://www.basno.com.

63. James D. Gordon III, "How Not to Succeed in Law School," *Yale Law Journal* 100 (1990–1991): 1679.

64. Eliot Freidson, "The Theory of Professions: State of the Art," in *The Sociology of the Professions,* ed. Robert Dingwall and Philip Lewis (New Orleans, LA: Quid Pro Books, 2014), n.1. See also Steve Kolo-wich, "Can Digital 'Badges' and 'Nanodegrees' Protect Job Seekers from a First-Round Knockout?," *Chronicle of Higher Education,* November 25, 2014, http://chronicle.com/article/Can-Digital-Badges -and/150257/. Staton, "The Degree Is Doomed."

65. Frederic Lardinois, "Duolingo Launches Its Certification Program to Take on TOEFL," *TechCrunch,* July 23, 2014, http://techcrunch

.com/2014/07/23/duolingo-launches-its-language-certification
-program/.

66. Erin Knight, "Open Badges for Lifelong Learning," White Paper, https://
wiki.mozilla.org/images/5/59/OpenBadges-Working-Paper_012312.pdf.

67. "Better Futures for 2 Million Americans through Open Badges,"
MacArthur Foundation press release, June 13, 2013, http://www
.macfound.org/press/press-releases/better-futures-2-million
-americans-through-open-badges/#sthash.5OjwBkFE.dpuf. See also
http://www.macfound.org/programs/digital-badges/.

68. Kaggle website, http://www.kaggle.com/solutions/connect.

69. Alireza Noruzi, "Google Scholar: The New Generation of Citation
Indexes," *International Journal of Libraries and Information Services*
55, no. 4 (2005): 169–235, http://www.librijournal.org/pdf
/2005-4pp170–180.pdf.

70. Michelene T. H. Chi, "Laboratory Methods for Assessing Experts' and
Novices' Knowledge," in *The Cambridge Handbook of Expertise and
Expert Performance,* ed. K. Anders Ericsson et al. (Cambridge, UK:
Cambridge University Press, 2006), 167.

71. J. Zhang, M. Ackerman, and L. Adamic, "Expertise Networks in
Online Communities: Structure and Algorithms," *Proceedings of the
16th International Conference on World Wide Web* (WWW'07) (New
York: ACM Press, 2007), 221–30.

72. M. R. Morris, J. Teevan, and K. Panovich, "A Comparison of Information Seeking Using Search Engines and Social Networks," in *Proceedings of the Fourth International Conference on Weblogs and Social
Media* (ICWSM '10) (Palo Alto, CA: AAAI Press, 2010), 291–94.

73. J. Weng et al., "Twitterrank: Finding Topic-Sensitive Influential
Twitterers," in *Proceedings of the ACM International Conference on
Web Search and Data Mining* (New York: Association for Computing
Machinery, 2010), 261–70.

74. Ha T. Tu and Johanna Lauer, "Word of Mouth and Physician Referrals
Still Drive Health Care Provider Choice," Center for Studying Health
System Change, December 2008, http://www.hschange.com
/CONTENT/1028/.

75. Fred Reichheld, *The Ultimate Question 2.0: How Net Promoter
Companies Thrive in a Customer-Driven World,* rev. and exp. ed.
(Cambridge, MA: Harvard Business Review Press, 2011).

76. Robert M. Bond et al., "A 61-Million-Person Experiment in Social Influence and Political Mobilization," *Nature* 489, no. 7415 (2012): 295–98; Yves-Alexandre de Montjoye et al., "The Strength of the Strongest Ties in Collaborative Problem Solving," *Scientific Reports* 4 (June 20, 2014), http://www.nature.com/srep/2014/140613/srep05277 /full/srep05277.html.

77. Eric von Hippel, Nikolaus Franke, and Reihnard Prügl, "'Pyramiding': Efficient Identification of Rare Subjects" (MIT Sloan School of Management Working Paper 4719–08, Cambridge, MA, October 2008).

78. Marion Poetz and Reinhard Prügel, "Find the Right Expert for Any Problem," *Harvard Business Review,* December 16, 2014, https://hbr .org/2014/12/find-the-right-expert-for-any-problem.

79. L. Sproull and S. Kiesler, "Reducing Social Context Cues: Electronic Mail in Organizational Communications," *Management Science* 32, no. 11 (1986): 1492–1512. See also Michael Fertik and David C. Thompson, *The Reputation Economy: How to Optimize Your Digital Footprint in a World Where Your Reputation Is Your Most Valuable Asset* (New York: Crown, 2015).

80. Lik Mui, Mojdeh Mohtashemi, and Ari Halberstadt, "A Computational Model of Trust and Reputation," *System Sciences, 2002. HICSS, Proceedings of the 35th Annual Hawaii International Conference on System Sciences* (IEEE, 2002), 2431–2439, http://www.hicss.hawaii.edu /HICSS_35/HICSSpapers/PDFdocuments/INIEB01.pdf.

81. See Peter Kollock, *The Production of Trust in Online Markets,* in *Advances in Group Processes,* vol. 16, ed. J. Lawler et al. (Greenwich, CT: JAI Press, 1999), 99. Separating the two sides of the transaction by time or space (such as purchasing something by mail or on credit) introduces greater risks: the party who moves second must be considered trustworthy or have some other form of guarantee. The formal infrastructure that exists to manage these risks is vast and includes such elements as credit card companies, credit rating services, public accounting firms, and—if the exchange goes bad— collection agencies or the court system.

82. Delia Ephron, "Ouch, My Personality, Reviewed," *New York Times,* August 29, 2014.

83. "A Brief History of LinkedIn," http://ourstory.linkedin.com/; "The Beginner's Guide to LinkedIn," Mashable, http://mashable.com/2012

/05/23/linkedin-beginners/; Gov. Ted Strickl, "I Tried to Live on Minimum Wage for a Week," *Politico,* http://www.politico.com /magazine/story/2014/07/a-mile-in-shoes-of-the-minimum-wage -worker-109418.html; "By the Numbers: 70 Amazing LinkedIn Statistics," February 9, 2014, http://expandedramblings.com/index .php/by-the-numbers-a-few-important-linkedin-stats/# .Uwp87vbNN2s.

84. Omar Alonso, Premkumar T. Devambu, and Michael Getz, "Expertise Identification and Visualization from CVS" (New York: ACM, 2004), http://citeseerx.ist.psu.edu/viewdoc/download?doi=10.1.1.163 .4643&rep=rep1&type=pdf.

85. Laura Dabbish et al., "Social Coding in GitHub: Transparency and Collaboration in an Open Software Repository" (Carnegie Mellon University, 2012), http://www.cs.cmu.edu/~xia/resources/Documents /cscw2012_Github-paper-FinalVersion-1.pdf; and Daniel Terdiman, "Forget LinkedIn: Companies Turn to Github to Find Tech Talent," *C/Net,* August 17, 2012, http://news.cnet.com/8301–10797_3 –57495099–235/forget-linkedin-companies-turn-to-github-to-find -tech-talent/.

86. Chrysthanthos Dellarocas, "The Digitization of Word of Mouth: Promise and Challenges of Online Feedback Mechanisms," *Management Science* 49, no.10 (2003): 1407–24, http://dl.acm.org/citation.cfm ?id=970709.

87. Kio Stark, *Don't Go Back to School: A Handbook for Learning Anything* (Kio Stark, 2013).

88. Stack Overflow, "Careers 2.0," http://careers.stackoverflow.com. See also Ashton Anderson et al., "Discovering Value from Community Activity on Focused Question Answering Sites: A Case Study of Stack Overflow," *Proceedings of the 18th ACM SIGKDD International Conference on Knowledge Discovery and Data Mining* (New York: ACM, 2012), 850–58, http://www.cs.cornell.edu/home/kleinber/kdd12 -qa.pdf.

89. See, e.g., P. Resnick et al., "Reputation Systems," *Communications of the ACM* 43, no. 12 (2000): 45–48.

90. N. Levina and Manuel Arriaga, "Distinction and Status Production on User-Generated Content Platforms: Using Bourdieu's Theory of Cultural Production to Understand Social Dynamics in Online

Fields," forthcoming in *Information Systems Research* 25, no. 3, 468–488, http://dx.doi.org/10.1287/isre.2014.0535.

91. "O*NET OnLine Help: OnLine Overview," *O*NET OnLine,* http://www.onetonline.org/help/online/. See also Aneesh Chopra and Nick Sinai, "Unlocking Labor and Skills Data for Americans: A Roundtable Discussion with Business and Policy Leaders" (Joan Shorenstein Center on Media, Politics, and Public Policy, Harvard Kennedy School, March 5, 2015), https://medium.com/@Shoren-steinCtr/unlocking-labor-and-skills-data-for-america-79a8dee43de8.

5. Experimenting with Smarter Governance

Epigraph: Susan L. Moffitt, *Making Policy Public: Participatory Bureaucracy In American Democracy* (Cambridge, UK: Cambridge University Press, 2014).

1. This chapter draws on an earlier published article: Beth Simone Noveck, "Crowdsourcing Wisely: Using Expert Networking to Enhance Medical Device Review at the Food and Drug Administration," *Yale Law and Policy Review* 32 (2014): 2.

2. Alok Das, "Transition: Locating and Integrating Members for Virtual Ad-hoc Teams" (Air Force SBIR/STTR Transition Story, July 2013), http://www.afsbirsttr.com/Publications/Documents/Transition -20130725-AF06-072-PeoplePoint.pdf.

3. W. Daniel Hillis, "Aristotle (The Knowledge Web)," *Edge,* May 6, 2004, http://edge.org/conversation/aristotle-the-knowledge-web.

4. Ibid.

5. Alton Hoover, "Aristotle—A Social Networking Solution Designed and Built for the Air Force Research Laboratory" (presented at the Aerospace and Electronics Conference, Dayton, Ohio, 2008), 36. See also The Defense Threat Reduction Agency, "The Defense Technical Information Center (DTIC): Aristotle Launched," August 6, 2010, http://www.dtra.mil/NewsMultimedia/NewsArticle9.aspx; and Open Government Initiative, "Aristotle" (blog), *The White House,* http://www.whitehouse.gov/open/innovations/Dod-Aristotle.

6. Cliff Kuang, "The Smart Problem-Solving behind Android's Awesome New Design Language | Design," *WIRED,* June 30, 2014, http://www

.wired.com/2014/06/the-big-ideas-behind-androids-awesome-new
-design-language/?mbid=social_fb.

7. Ron Kohavi et al., "Online Controlled Experiments at Large Scale," *Microsoft Research* (2013), http://www.exp-platform.com/Documents /2013%20controlledExperimentsAtScale.pdf; Ron Kohavi, Roger Longbotham, and Toby Walker, "Online Experiments: Practical Lessons," in *IEEE Computer Society: Web Technologies* (September 2010), http://www.exp-platform.com/Documents/IEEE2010ExP.pdf.

8. Jaron Lanier, "Jaron Lanier on Lack of Transparency in Facebook Study," *New York Times,* June 30, 2014, http://www.nytimes.com/2014 /07/01/opinion/jaron-lanier-on-lack-of-transparency-in-facebook -study.html. Duncan Watts, "Stop Complaining about the Facebook Study. It's a Golden Age for Research," *The Guardian,* July 7, 2014, http://www.theguardian.com/commentisfree/2014/jul/07/facebook -study-science-experiment-research.

9. Elizabeth Gudrais, "Rebooting Social Science," *Harvard Magazine* 116, no. 6 (August 2014): 54.

10. Jim Manzi, *Uncontrolled: The Surprising Payoff of Trial-and-Error for Business, Politics, and Society* (New York: Basic Books, 2012), xii.

11. Laura Haynes et al., "Test, Learn, Adapt: Developing Public Policy with Randomized Controlled Trials," Cabinet Office Behavioural Insights Team, https://www.gov.uk/government/uploads/system /uploads/attachment_data/file/62529/TLA-1906126.pdf.

12. "There's a Strong Personal Connection" (blog), *Sage Bionetworks,* April 9, 2015, http://sagebase.org/theres-a-strong-personal-connection/.

13. Elinor Ostrom, *Understanding Institutional Diversity* (Princeton, NJ: Princeton University Press, 2005), 97.

14. Edward Castronova, "A Test of the Law of Demand in a Virtual World: Exploring the Petri Dish Approach to Social Science" (CESifo Working Paper Series no. 2355, July 2008). http://papers.ssrn.com/sol3 /papers.cfm?abstract_id=1173642.

15. Duncan Watts, "Computational Social Sciences: Exciting Progress and Future Directions," *The Bridge* (Winter 2013), https://www.nae .edu/File.aspx?id=106114.

16. David Lazer et al., "Computational Social Science," *Science* 326 (2009): 721–23, http://www.ncbi.nlm.nih.gov/pubmed/19197046.

17. Laura Haynes et al., "Test, Learn, Adapt: Developing Public Policy with Randomized Controlled Trials" (Cabinet Office Behavioural Insights Team): 13–14, https://www.gov.uk/government/uploads /system/uploads/attachment_data/file/62529/TLA-1906126.pdf.

18. Cass Sunstein and Richard Thaler, *Nudge: Improving Decisions about Health, Wealth, and Happiness* (New York: Penguin, 2009).

19. Haynes, "Test, Learn, Adapt," 12.

20. "Judah Folkman: An Inspiration for the Tobin Project," http://www .tobinproject.org/about/judah-folkman. See also Jonathan West, Mona Ashiya, and Ambuj Sagar, "Judah Folkman and the War on Cancer," *Harvard Business Review*, June 4, 2004, http://hbr.org /product/judah-folkman-and-the-war-on-cancer/an/604091-PDF -ENG.

21. "If Only We Knew" (blog), *The GovLab*, http://thegovlab.org/if-only -we-knew/.

22. Geoff Mulgan, "True Collective Intelligence: A Sketch of a Possible New Field," *Philosophy and Technology* 27, no. 1 (2014):133–42, http://philpapers.org/rec/MULTCI-3; Clive Thomson, *Smarter Than You Think: How Technology Is Changing Our Minds for the Better* (New York: Penguin, 2013).

23. Duncan Watts, "Computational Social Science: Exciting Progress and Future Directions," n.d., https://www.nae.edu/Publications/Bridge /106112/106118.aspx.

24. Tiago Peixoto and Fredrik Sjoberg, "The Haves and Have Nots: Is Civic Tech Impacting the People Who Need It Most?" World Bank Group, https://www.theengineroom.org/wp-content/uploads/The -haves-and-the-have-nots.pdf; and Fredrik M. Sjoberg, Jonathan Mellon, and Tiago Peixoto, "The Effect of Government Responsive-ness on Future Political Participation," World Bank Group, Feb-ruary 26, 2015, https://www.mysociety.org/files/2015/03/SSRN -id2570898.pdf.

25. See Clifford Geertz, "Thick Description: Toward an Interpretive Theory of Culture," in *The Interpretation of Cultures: Selected Essays* (New York: Basic Books), 3–30.

26. Open Government Partnership, "National Action Plan for the United States of America," *The White House*, September 20, 2011, http://www .whitehouse.gov/sites/default/files/us_national_action_plan_final_2

.pdf; Open Government Partnership, "Second Open Government National Action Plan for the United States of America," *The White House,* December 15, 2013, http://www.whitehouse.gov/sites/default /files/docs/us_national_action_plan_6p.pdf.

27. Memorandum on Transparency and Open Government from President Barack Obama to the Heads of Executive Departments and Agencies, January 21, 2009, http://www.whitehouse.gov/the_press _office/TransparencyandOpenGovernment.

28. "ExpertNet: Getting Started," http://expertnet.wikispaces.com /Getting+Started.

29. Ibid.

30. "Original Draft, ExpertNet," http://expertnet.wikispaces.com /Original+Draft.

31. ExpertNet was withdrawn from the Open Government Partnership National Action Plan because of difficulties in implementation and conflicts with the Federal Advisory Committee Act (FACA) and, according to the government self-assessment, the existence of private sector platforms to enable experts to make proposals. Independent Reporting Mechanism, Open Government Partnership, United States Progress Report 2011–2013, 10, http://www.opengovpartnership.org /sites/default/files/IRM100513-USA-Report-LIVE.pdf.

32. *Food and Drug Administration Safety and Innovation Act,* Public Law 112–144,*Statute* 993 (2012), 126 codified at *U.S. Code* 21 in various sections.

33. Ariel Dora Stern, "Innovation under Regulatory Uncertainty: Evidence from Medical Technology," Harvard Business School, July 15, 2014.

34. Diana M. Zuckerman, Paul Brown, and Steven E. Nissen, "Medical Device Recalls and the FDA Approval Process," *Archives of Internal Medicine* 171 (2011): 1006.

35. Elaine Quijano, "Report: Medical Implants Rarely Tested," *CBS News,* March 28, 2012, http://www.cbsnews.com/8301–505269_162–57405666 /report-medical-implants-rarely-tested.

36. Greg Farrell and David Voreacos, "J&J Hip Failure Rate as High as 49 Percent, U.K. Doctors Say," *Bloomberg News,* March 9, 2011, http://www.bloomberg.com/news/2011–03–09/j-j-hip-replacement -failure-rate-may-be-49-u-k-orthopedists-group-says.html.

37. Diana M. Zuckerman, Paul Brown, Steven E. Nissen, "Medical Device Recalls and the FDA Approval Process," *Archives of Internal Medicine* (February 14, 2011), http://doi: 10.2002/archintenmedia.201.30.

38. Andrew Pollack, "Medical Treatment, Out of Reach," *New York Times*, February 9, 2011, http://www.nytimes.com/2011/02/10/business /10device.html.

39. Ariel D. Stern et al., "Predicting Adverse Event Reports with Regulatory Approval Data: The Case of New Cardiovascular Devices," forthcoming.

40. For more on premarket medical device review, see "PMA Approval Process," http://1.usa.gov/175WI65.

41. Jessica Nadine Hernandez, Office of Science and Engineering Laboratories, Food and Drug Administration, 2013, unpublished draft blog posting, on file with author.

42. Office of Device Evaluation, *Annual Performance Report: Fiscal Year 2009*, U.S. Food and Drug Administration 4 (2009), http://www.fda .gov/downloads/AboutFDA/CentersOffices/OfficeofMedicalProductsa ndTobacco/CDRH/CDRHReports/UCM223893.pdf.

43. Susan Jaffe, "FDA Reforms Hampered by Budget Freeze," *Lancet* 380 (2012): 1458, 1458–59.

44. U.S. Food and Drug Administration, "Innovation Pathway," http://www.fda.gov/AboutFDA/CentersOffices/OfficeofMedicalProdu ctsandTobacco/CDRH/CDRHInnovation/InnovationPathway/default .htm.

45. U.S. Food and Drug Administration, "CDRH Network of Experts," http://www.fda.gov/AboutFDA/CentersOffices/OfficeofMedicalProdu ctsandTobacco/CDRH/ucm289534.htm.

46. Joanna Brougher et al., "A Practical Guide to Navigating the Medical Device Industry: Advice from Experts in Industry," *Intellectual Property and Academia* (2011): 32.

47. A Special Government Employee (SGE) is one who works for a term "not to exceed one hundred and thirty days during any period of three hundred and sixty-five consecutive days." See *U.S. Code* 18 (2012), § 202.

48. U.S. Food and Drug Administration, "FDA Outlines Plans for an Outside Network of Scientific Experts," October 4, 2011, http://www

.fda.gov/NewsEvents/Newsroom/PressAnnouncements/ucm274351
.htm.

49. U.S. Food and Drug Administration, "CDRH Network of Experts."

50. U.S. Food and Drug Administration, *Improvements in Device Review: Results of CDRH's Plan of Action for Premarket Review of Devices* (November 2012), http://www.fda.gov/downloads/AboutFDA /ReportsManualsForms/Reports/UCM329702.pdf.

51. *Competitiveness and Regulation: The FDA and the Future of America's Biomedical Industry* (La Jolla: California Healthcare Institute, February 2011), iii, http://www.bcg.com/documents/file72060.pdf.

52. Fenwick and West Life Sciences Group, *The Regulatory Future of mHealth: FCC, FDA, and the United States Congress, JDSupra Business Advisor* (June 19, 2012), http://www.jdsupra.com/legalnews/the -regulatory-future-of-mhealth-fcc-f-22552/.

53. Memorandum of Understanding between the Federal Communications Commission and the Food and Drug Administration Center for Devices and Radiological Health, July 2010, http://www.elsevierbi .com/~/media/Images/Publications/Archive/The%20Gray%20Sheet /36/031/01360310001/080210_fda_fcc_mou.pdf.

54. Letter from Marsha Blackburn et al., Members, U.S. Congress, to Margaret Hamburg, Commissioner, U.S. Food and Drug Administration, and Julius Genachowski, Chairman, U.S. Federal Communication Commission, April 3, 2012, http://blackburn.house.gov /uploadedfiles/letter_from_congress_to_fda_and_fcc_-_3apr2012 .pdf; see also Letter from Bradley Merrill Thompson of Epstein Becker & Green, P.C., to U.S. Food and Drug Administration, October 11, 2012, http://mhealthregulatorycoalition.org/wp-content/uploads/2010 /06/MRC-Comment-Letter-on-UDI-Proposal-10112012.pdf.

55. *Food and Drug Administration Safety and Innovation Act,* Public Law 112–144, Sec. 618, 21 USC § 301 (2012).

56. U.S. Food and Drug Administration, "CDRH Network of Experts."

57. U.S. Food and Drug Administration, "Distribution of Full-Time Equivalent Employment Program Level," http://www.fda.gov /downloads/AboutFDA/ReportsManualsForms/Reports /BudgetReports/UCM301553.pdf; See National Institutes of Health, "About NIH," http://www.nih.gov/about/; U.S. Department of

Health and Human Services, *2011 Strategic Sustainability Perfor-mance Plan* (2011), 4, http://www.hhs.gov/about/sustainability /2011plan_summary.pdf.

58. U.S. Food and Drug Administration, "Advisory Committees," http://www.fda.gov/AdvisoryCommittees /CommitteesMeetingMaterials/MedicalDevices /MedicalDevicesAdvisoryCommittee/.

59. Harvard Catalyst, "Open Source Community," http://profiles.catalyst .harvard.edu/?pg=community.

60. Ibid.

61. Jessica N. Hernandez, FDA Profiles team member, telephone inter-view with the author, March 31, 2014.

62. Anjali Kataria, Food and Drug Administration, Entrepreneur in Residence, and Jessica N. Hernandez, Program Analyst, Office of Science and Engineering Laboratories, Food and Drug Administra-tion (presentation on the OSEL Research Networking Software Pilot, Silver Spring, MD, October 1, 2013).

63. Dean Krafft et al., "Enabling National Networking of Scientists," (presentation at the 2010 Web Science conference, Raleigh, NC, April 2010, http://journal.webscience.org/316/2/websci10_submission _82.pdf.

64. VIVO, "About VIVO," http://www.vivoweb.org/about.

65. See VIVO, "Semantic Modeling of Scientist: The VIVO Ontology," http://vivoweb.org/files/SemanticModeling.pdf.

66. Harvard Catalyst, "About Harvard Catalyst Profiles," http://connects .catalyst.harvard.edu/Profiles/about/default.aspx.

67. UCSF Open Proposals, "About UCSF Open Proposals," https://open -proposals.ucsf.edu/.

68. Eva Guinan, Kevin J. Boudreau and Karim R. Lakhani, "Experiments in Open Innovation at Harvard Medical School," *MIT Sloan Manage-ment Review* 54 (2013), 45, 46.

69. Harvard Catalyst, "Congratulations to the Winners of the First Harvard Catalyst/InnoCentive Ideation Challenge: 'What Do We Not Know to Cure Type 1 Diabetes?'" September 29, 2010, http://catalyst .harvard.edu/news/news.html?p=1966.

70. Karim Lakhani, interview with the author, quoted in Beth Simone Noveck, "Innovating the Innovation Process: Re-imagining University

Research" (blog), *The GovLab,* June 5, 2013, http://thegovlab.org
/innovating-the-innovation-process-re-imagining-university-research/.

71. Guinan et al., *Experiments in Open Innovation,* 50.

72. Ibid., 47–48.

73. Jianguo Liu et al., "Ecological Degradation in Protected Areas: The
Case of Wolong Nature Reserve for Giant Pandas," *Science* 292 (2001):
98.

74. Barbara R. Jasny, "Governance by the People," *Science* 314 (2013): 11.

75. Jason Miller and Tom Kalil, "Crowdsourcing Ideas to Accelerate
Economic Growth and Prosperity through a Strategy for American
Innovation," July 28, 2014, http://www.whitehouse.gov/blog/2014/07
/28/crowdsourcing-ideas-accelerate-economic-growth-and-prosperity
-through-strategy-ameri.

76. Full text of the Magna Carta for Philippine Internet Freedom
(MCPIF), http://democracy.net.ph/full-text/. Jessica McKenzie,
"Crowdsourced Internet Freedom Bill a First for Filipino Lawmakers,"
TechPresident, July 31, 2013, http://techpresident.com/news/wegov
/24226/crowdsourced-internet-freedom-bill-first-philippine
-lawmakers; "Marco Civil da Internet" (blog), Rio de Janeiro Law
School (FGV), May 9, 2014, http://direitorio.fgv.br/noticia/the
-brazilian-civil-rights-framework-for-the-internet.

77. Henry Chesbrough, *Open Innovation: The New Imperative for
Creating and Profiting from Technology* (Boston, MA: Harvard
Business School Press, 2003).

78. Youth@Work Bhutan (2014), https://communityplanit.org
/youthbhutan/.

79. See, for example, Bo Pang and Lillian Lee, "Opinion Mining and
Sentiment Analysis," *Foundations and Trends in Information Retrieval*
2.1–2.2 (2008): 1–135, http://bit.ly/UaCBwD.

80. Ravi Arunachalam and Sandipan Sarkar, "The New Eye of Government: Citizen Sentiment Analysis in Social Media" (IJCNLP 2013
Workshop on Natural Language Processing for Social Media [So-
cialNLP], Nagoya, Japan, October 14, 2014), 23–28, https://www
.aclweb.org/anthology/W/W13/W13–42.pdf#page=35.

81. Jolie O'Dell, "Could Twitter Data Replace Opinion Polls? [STUDY],"
Mashable, May 11, 2010, http://mashable.com/2010/05/11/twitter-data
-opinion-polls/.

82. Rob Procter, "How 2.6m Tweets Were Analysed to Understand Reaction to the Riots," *The Guardian,* December 7, 2011, http://www .theguardian.com/uk/2011/dec/07/how-tweets-analysed-understand -riots.

83. Rebecca Chao, "The Secret to That Potato-Salad Kickstarter Campaign's Success," *The Atlantic,* July 24, 2014, http://www.theatlantic .com/technology/archive/2014/07/the-data-behind-that-potato-salad -kickstarter/374998/.

84. *Spacehive,* https://spacehive.com/.

85. *Neighbor.ly,* https://neighbor.ly/.

86. Elizabeth M. Gerber and Julie Hui, "Crowdfunding: Motivations and Deterrents for Participation," http://egerber.mech.northwestern.edu /wp-content/uploads/2012/11/Gerber_Crowdfunding _MotivationsandDeterrents.pdf.

87. *TED,* "TED Open Translation Project," http://www.ted.com/about /programs-initiatives/ted-open-translation-project.

88. Carsten Eickhoff and Arjen P. de Vries, "Increasing Cheat Robustness of Crowdsourcing Tasks," *Information Retrieval* 16, no. 2 (April 2013), http://link.springer.com/article/10.1007 /s10791–011–9181–9.

89. See, for example, Geoffrey Barbier et al., "Maximizing Benefits from Crowdsourced Data," *Computational and Mathematical Organization Theory* 18, no. 3 (September 2012): 2–23, http://www.public.asu.edu /~hga016/papers/CMOT.pdf.

90. City of Boston, "Street Bump: Help Improve Your Streets," http://www .cityofboston.gov/doit/apps/streetbump.asp.

91. Anna Scott, "Open data: How Mobile Phones Saved Bananas from Bacterial Wilt in Uganda," *The Guardian,* February 11, 2015, http://www.theguardian.com/global-development-professionals -network/2015/feb/11/open-data-how-mobile-phones-saved-bananas -from-bacterial-wilt-in-uganda.

92. United Nations Institute for Training and Research, "UN-OSAT Uses Crowd Source App to Involve Locals in Monitoring Floods in Bangkok," November 8, 2011, http://www.unitar.org /unosat-uses-crowd-source-app-involve-locals-monitoring-floods -bangkok.

93. Sebastian Deterding, "Gamification: Designing for Motivation," *Social Mediator* (July/August 2012), https://www.cs.auckland.ac.nz/courses/compsci747s2c/lectures/paul/p14-deterding.pdf.

94. Daren Brabham, "Using Crowdsourcing in Government," IBM Center for the Business of Government, 2013, http://www.businessofgovernment.org/sites/default/files/Using%20Crowd sourcing%20In%20Government.pdf.

95. S. L. Bryant, A. Forte, and A. Bruckman, "Becoming Wikipedian: Transformation of Participation in a Collaborative Online Encyclopedia," in *Proceedings of the 2005 International ACM SIGGROUP Conference on Supporting Group Work*, 1–10, http://www.cc.gatech.edu/~asb/papers/bryant-forte-bruckman -group05.pdf.

96. Jeff Howe, *Crowdsourcing: How the Power of the Crowd Is Driving the Future of Business* (New York: Crown Business, 2009), 288.

97. Johann Füller, Katja Hutter, and Mirijam Fries, "Crowdsourcing for Goodness Sake: Impact of Incentive Preference on Contribution Behavior for Social Innovation," *Advances in International Marketing* 23 (2012), http://www.emeraldinsight.com/books.htm?chapterid =17049532.

98. Nicolas Kaufmann et al., "More Than Fun and Money: Worker Motivation in Crowdsourcing—A Study on Mechanical Turk," 2011, http://www.researchgate.net/publication/220894276_More_than_fun _and_money._Worker_Motivation_in_Crowdsourcing_-_A_Study _on_Mechanical_Turk/file/9fcfd50e5afe007d78.pdf; Stacey Kuznetsov, "Motivations of Contributors to Wikipedia," *ACM SIGCAS Com-puters and Society Newsletter,* June 2006, 3–4.

99. Lena Mamykina et al., "Design Lessons from the Fastest Q&A Site in the West," *Proceedings of SIGCHI Conference on Human Factors in Computing Systems* (Vancouver, BC: Association for Computing Machinery, May 7–12, 2011), 2857, 2865.

100. Stacey Kuznetsov, "Motivations of Contributors to Wikipedia," 3.

101. Rishab Aiyer Ghosh, "Understanding Free Software Developers: Findings from the FLOSS Study," in *Perspectives on Free and Open Software,* ed. Joseph Feller et al. (Cambridge, MA: MIT Press, 2007), 23, 35.

102. Mamykina et al., "Design Lessons," 2857, 2865.

103. Ashton Anderson et al., "Steering User Behavior with Badges," *2013 Proceedings of the 22nd International Conference on World Wide Web* (International World Wide Web Conferences Steering Committee, Geneva, Switzerland, 2013): 95–106.

104. Al M. Rashid et al., "Motivating Participation by Displaying the Value of Contribution," *Proceedings of the SIGCHI Conference on Human Factors in Computing Systems* (Montreal: CHI, 2006), 955.

105. Cliff Lampe and Erik Johnston, "Follow the (Slash) Dot: Effects of Feedback on New Members in an Online Community," *Proceedings of the 2005 International ACM SIGGROUP Conference on Supporting Group Work* (New York: ACM, 2005), 11–20.

106. Charles Leadbeater, "Hooked on Labs," in *The Long + Short,* http://thelongandshort.org/issues/season-two/hooked-on-labs.html.

107. Pelle Ehn, Elisabet M. Nilsson, and Richard Topgaard, eds., *Making Futures* (Cambridge, MA: MIT Press, 2014); Zaid Hassan, *The Social Labs Revolution: A New Approach to Solving Our Most Complex Challenges,* 1st ed. (San Francisco, CA: Berrett-Koehler, 2014); Ruth Pottick, Peter Baeck, and Philip Colligan, "i-Teams: The Teams and Funds Making Innovation in Governments Happen around the World," NESTA, 2014, http://www.nesta.org.uk/sites/default/files/i-teams_june_2014.pdf.

108. Guerogi Kossinets and Duncan Watts, "Empirical Analysis of an Evolving Social Network," *Science* (January 2006): 88–90; Lars Backstrom, Dan Huttenlocher, and Jon Kleinberg, "Group Formation in Large Social Networks: Membership, Growth, and Evolution," *Proceedings of the 12th ACM SIGKDD International Conference on Knowledge Discovery and Data Mining* (New York: ACM, 2006): 44–54, http://wiki.cs.columbia.edu/download/attachments/1979/Group+Formation+in+Large+Social+Networks-backstrom.pdf; Eytan Bakshy, Brian Karrer, and Lada A. Adamic, "Social Influence and the Diffusion of User Created Content," *Proceedings of the 10th ACM Conference on Electronic Commerce* (New York: ACM, 2009): 325–34, http://dl.acm.org/citation.cfm?id=1566421; Michael S. Bernstein et al., "Quantifying the Invisible Audience in Social Networks," *Proceedings of the SIGCHI Conference on Human*

Factors in Computing Systems, (New York: ACM, 2013): 21–30, http://hci.stanford.edu/publications/2013/invisibleaudience /invisibleaudience.pdf.

109. Duncan Watts, "Computational Social Science: Exciting Progress and Future Directions," April 7, 2014, http://www.nap.edu/openbook.php ?record_id=18558&page=17.

110. Joe Raelin, "Seeking Conceptual Clarity in the Action Modalities," *Action Learning: Research and Practice* 6, no. 1 (2009): 17–24, http://www.informaworld.com.

111. Michelle Andrews et al., "Mobile Ad Effectiveness: Hyper-Contextual Targeting with Crowdedness," December 11, 2014, *Marketing Science,* forthcoming, http://dx.doi.org/10.2139/ssrn.2439388.

112. Dominik Molitor et al., "The Impact of Smartphones, Bar-code Scanning, and Location-based Services on Consumers' Search Behavior" (Proceedings of the 2013 International Conference on Information Systems [ICIS], Milano, Italy, October 2013), http://aisel.aisnet.org/icis2013/proceedings /ResearchInProgress/87/.

113. Mark Moore, *Creating Public Value: Strategic Management in Government* (Cambridge, MA: Harvard University Press, 1995), 20.

6. Why Smarter Governance May Be Illegal

Epigraph: Theodore Roosevelt, "The Right of the People to Rule," Carnegie Hall, New York (March 12, 1912), http://teachingamericanhistory.org /library/document/the-rights-of-the-people-to-rule/.

1. See letter from Thomas Jefferson to James Hutchinson, March 12, 1791, in *The Papers of Thomas Jefferson,* vol. 19, ed. Julian P. Boyd (Princeton, NJ: Princeton University Press, 1974), 614; Wendy Ginsberg, "Federal Advisory Committees: An Overview," Congressional Research Service, April 16, 2009, 2, http://fpc.state.gov /documents/organization/122888.pdf.

2. William R. Funk, "Public Participation and Transparency in Administrative Law—Three Examples as an Object Lesson," *Administrative Law Review* 61 (2009): 172.

3. See Beth Simone Noveck, "The Electronic Revolution in Rulemaking," *Emory Law Review* 53 (2004): 433.

4. Horowitz and Kamvar discuss the superiority of chat-like interfaces for honest conversations. D. Horowitz and S. Kamvar, "The Anatomy of a Large-scale Social Search Engine," in *Proceedings of the International World Wide Web Conference* (WWW '10) (New York: ACM, 2010), 431–40.

5. James T. Reilly, "Federal Advisory Committee Act: Inhibiting Effects upon the Utilization of New Media in Collaborative Governance and Agency Policy Formation," Report to the Administrative Conference, Draft, March 17, 2011.

6. *EPA Science Advisory Board Reform Act of 2014,* HR 1422, 113th Cong. (2013–2014), https://www.congress.gov/bill/113th-congress /house-bill/1422.

7. Ginsberg, "Federal Advisory Committees."

8. Steven P. Croley and William T. Funk, "The Federal Advisory Committee Act and Good Government," *Yale Journal on Regulation* 14 (1997): 460. See also U.S. Congress, House of Representatives, HR 85–576, "Amending the Administrative Expenses Act of 1946, and for Other Purposes," 85th Cong., 1st sess. (June 17, 1957), 2.

9. Funk, "Public Participation and Transparency," 183; and Croley and Funk, "Federal Advisory Committee Act," 463.

10. James Conant, *Science and Common Sense* (New Haven, CT: Yale University Press, 1951), 337.

11. Wendy R. Ginsberg, *Advisory Committees: A Primer* (Washington, DC: U.S. Congressional Research Service, 2010), 3.

12. U.S. Congress, House Committee on the Judiciary, Antitrust Subcommittee (Subcommittee no. 5), *WOC's [Without Compensation Government Employees] and Government Advisory Groups,* Hearings, 84th Cong., 1st sess. (August 4, 1955), S. Hrg. Part 1, 586–87.

13. Funk, "Public Participation and Transparency," 177.

14. Charles H. Koch Jr., "James M. Landis, The Administrative Process," *Faculty Publications,* Paper 633 (1996), http://scholarship.law.wm.edu /facpubs/633.

15. Conant, *Science and Common Sense,* 337.

16. Jerry W. Markham, "The Federal Advisory Committee Act," *University of Pittsburgh Law Review* 35, no. 3 (1974): 558–59.

17. Ginsburg, "Federal Advisory Committees," 3.
18. Individual views of Hon. John E. Moss, Federal Advisory Committee Standards Act Legislative History, Committee on Government Operations, Report no. 92-1017 (April 25, 1971), 18, https://bulk .resource.org/gao.gov/92-463/00005E10.pdf.
19. Full text of "Presidential Commissions," Hearings before the Senate Subcommittee on Administrative Practice and Procedure 92–1 (May 25, 26; June 22, 23; July 14 and 27, 1971), 3, cited in Paul Light, *Government by Investigation: Congress, Presidents, and the Search for Answers 1945–2012* (Washington, DC: Brookings Institution Press, 2014), 214.
20. The Economic Opportunity Act mandated maximum feasible participation among the poor in community action programs.
21. Mark Brown, *Science in Democracy: Expertise, Institution, and Representation* (Cambridge, MA: MIT Press, 2009), 96.
22. Steven P. Croley, "Practical Guidance on the Applicability of the Federal Advisory Committee Act," *Administrative Law Journal* 10 (1996): 141.
23. Harold Wilensky, *Organizational Intelligence* (New York: Basic Books, 1969), 169–72.
24. The Negotiated Rulemaking Act was reauthorized in 1996 and is not incorporated into the Administrative Procedure Act. See *Administrative Procedure Act of 1946*, §§ 561–700 (1946); David M. Pritzker and Deborah S. Dalton, *Negotiated Rulemaking Sourcebook* (Washington, DC: Office of the Chairman, Administrative Conference of the U.S., 1990); and Cary Coglianese, Heather Kilmartin, and Evan Mendelson, "Transparency and Public Participation in the Federal Rulemaking Process: Recommendations for the New Administration," *George Washington Law Review* 77 (1997): 924–972.
25. Richard H. Pildes and Cass R. Sunstein, "Reinventing the Regulatory State," *University of Chicago Law Review* 62 (1995): 1ff. See also Noveck, "The Electronic Revolution," 433ff.
26. William R. Funk, "The Paperwork Reduction Act: Paperwork Reduction Meets Administrative Law," *Harvard Journal on Legislation* 24 (1987): 93.
27. Pildes and Sunstein, "Reinventing the Regulatory State," 4.
28. *Paperwork Reduction Act of 1980, U.S. Code* 44 (1980), § 3501(4).

29. Ibid., § 3502 (3)(A). Office of Management and Budget, "Memorandum: Social Media, Web-Based Interactive Technologies"; and *Paperwork Reduction Act,* 2, http://www.whitehouse.gov/sites/default /files/omb/assets/inforeg/SocialMediaGuidance_04072010.pdf.

30. Office of Management and Budget, "Memorandum: New Fast-Track Process for Collecting Service Delivery Feedback under the Paperwork Reduction Act," June 15, 2011, http://www.whitehouse.gov/sites /default/files/omb/memoranda/2011/m11-26.pdf.

31. *Paperwork Reduction Act of 1980,* § 3503(c)(3).

32. For an inventory of currently approved collections, with OMB control numbers, see http://www.reginfo.gov.

33. Office of Management and Budget, "Memorandum: Social Media."

34. Stack Overflow, "How to Ask," http://stackoverflow.com/questions/ask /advice?.

35. Office of the President, "Regulatory Planning and Review," Executive Order 12866, September 30, 1993, http://www.reginfo.gov/public/jsp /Utilities/EO_12866.pdf.

36. Funk, "Public Participation and Transparency," 182.

37. Chris Mooney, *The Republican War on Science* (New York: Basic Books, 2005).

38. Office of the President, "Termination and Elimination of Federal Advisory Committees," Executive Order 12838, February 10, 1993, http://www.gpo.gov/fdsys/pkg/WCPD-1993-02-15/pdf/WCPD-1993-02 -15-Pg166.pdf.

39. Office of Management and Budget, "Management of Federal Advisory Committees," Circular A-135, October 5, 1994, https://www .whitehouse.gov/omb/circulars_a135/.

40. Kathleen Doherty, "Seeking Experts or Agents of Control: The Use of Advisory Committees in Bureaucratic Policymaking," September 3, 2013, http://www.vanderbilt.edu/csdi/events/KMD-AdComs_V.pdf.

41. U.S. General Services Administration, *The Federal Advisory Committee Act* (FACA) brochure, http://www.gsa.gov/portal/content /101010.

42. Julia Metz, "EU Commission Expert Groups: Between Inclusive and Effective Policy-making," in *Expertise and Democracy,* ed. Cathrine Holst, ARENA Report 1/14 (Arena Centre for European Studies, University of Oslo, Norway, February 2014), http://www.sv.uio.no

/arena/english/research/publications/arena-publications/reports/2014
/report-01–14.pdf.

43. Ibid., 264.

44. John R. Moodie and Cathrine Holst, "For the Sake of Democracy? The
European Commission's Justifications for Democratising Expertise,"
in Holst, *Expertise and Democracy,* 299.

45. Ibid.

46. Sheila Jasanoff, "Quality Control and Peer Review in Advisory
Science," in *The Politics of Scientific Advice: Institutional Design for
Quality Assurance,* ed. Justus Lentsch and Peter Weingart (Cam-
bridge, UK: Cambridge University Press), 29; See also Sheila Jasanoff,
The Fifth Branch: Science Advisers as Policymakers (Cambridge, MA:
Harvard University Press, 1990); Sheila Jasanoff, *Designs on Nature:
Science and Democracy in Europe and the United States* (Princeton,
NJ: Princeton University Press, 2005).

47. Ginsberg, "Federal Advisory Committees," 12, citing Amy B. Zegart,
"Blue Ribbons, Black Boxes: Toward a Better Understanding of Presidential Commissions," *Political Studies Quarterly* 34, no. 2 (June 2004): 372.

48. Julia Ryan, "American Schools vs. the World: Expensive, Unequal,
Bad at Math," *The Atlantic,* December 3, 2013, http://www.theatlantic
.com/education/archive/2013/12/american-schools-vs-the-world
-expensive-unequal-bad-at-math/281983; Pearson, *Index of Cognitive
Skills and Educational Attainment,* http://thelearningcurve.pearson
.com/index/index-ranking.

49. Organization for Economic Co-operation and Development, *United
States—Country Note—Education at a Glance: OECD indicators 2012*
(Paris: Organization for Economic Co-operation and Development,
2012), http://www.oecd.org/unitedstates/CN%20-%20United%20
States.pdf.

50. Robert Archibald and David Feldman, *Why Does College Cost So
Much?* (Oxford, UK: Oxford University Press, 2010); "The Future of
Universities: The Digital Degree," *The Economist,* June 28, 2014,
http://www.economist.com/news/briefing/21605899-staid-higher
-education-business-about-experience-welcome-earthquake-digital
?zid=316&ah=2f6fb672faf113fdd3b11cd1b1bf8a77.

51. The White House, Office of the Press Secretary, "Presidential Memorandum: Helping Struggling Federal Student Loan Borrowers Manage

Their Debt," June 9, 2014, http://www.whitehouse.gov/the-press-office /2014/06/09/presidential-memorandum-federal-student-loan -repayments. ("Over the past three decades, the average tuition at a public four-year college has more than tripled, while a typical family's income has increased only modestly. More students than ever are relying on loans to pay for college. Today, 71 percent of those earning a bachelor's degree graduate with debt, which averages $29,400.")

52. David Soo, "How Can the Department of Education Increase Innovation, Transparency and Access to Data?" (blog) *Home-room,* April 2014, http://www.ed.gov/blog/2014/04/how-can-the -department-of-education-increase-innovation-transparency-and -access-to-data.

53. Whitehouse.gov, "Higher Education," http://www.whitehouse.gov /issues/education/higher-education.

54. Ryan, "American Schools vs. the World."

55. Whitehouse.gov, "Reform for the Future," http://www.whitehouse.gov /issues/education/reform.

56. American Association of University Professors, "Teaching Millions or Making Millions?" http://www.aaup.org/news/teaching-millions-or -making-millions.

57. Clay Shirky, "The End of Higher Education's Golden Age" (blog) *Clay Shirky,* January 29, 2014, http://www.shirky.com/weblog/2014 /01/there-isnt-enough-money-to-keep-educating-adults-the-way -were-doing-it/.

58. Croley and Funk, "Federal Advisory Committee Act," 451.

59. Jeremy Shapiro, "Who Influences Whom? Reflections on U.S. Government Outreach to Think Tanks" (blog) *Brookings,* June 4, 2014, http://www.brookings.edu/blogs/up-front/posts/2014/06/04-us -government-outreach-think-tanks-shapiro.

60. Emily Rydel Ackman, "Getting Scholarship into Policy: Lessons from University-Based Bipartisan Scholarship Brokers" (Ph.D. diss., Arizona State University, 2013).

61. U.S. Department of Education "Who Serves on Committees" GSA .gov (2012), http://www.gsa.gov/portal/content/249045.

62. U.S. Department of Education, "About ACFSA," http://www2.ed.gov /about/bdscomm/list/acsfa/edlite-about.html.

63. Ibid.

64. U.S. Department of Education, "Advisory Committee Publications," http://www2.ed.gov/about/bdscomm/list/acsfa/edlite-publications .html.

65. Croley and Funk, "Federal Advisory Committee Act," 453.

66. Beth Simone Noveck, *Wiki Government* (Washington, DC: Brookings Institution Press, 2009), 27.

67. Union of Concerned Scientists, "Restoring Scientific Integrity in Policy-making: Scientists Sign-on Statement," 2008, http://www .ucsusa.org/our-work/center-science-and-democracy/promoting -scientific-integrity/scientists-sign-on-statement.html# .VVA6mdpViko.

68. Markham, *Federal Advisory Committee Act,* 566.

69. Carolyn Bingham Kello, "Note: Drawing the Curtain on Open Government: In Defense of the Federal Advisory Committee Act," *Brooklyn Law Review* 69 (2002–2003): 344.

70. Federal Acquisition Regulations, 48 CFR 35.017 —Federally Funded Research and Development Centers, http://www.law.cornell.edu/cfr /text/48/35.017. Jill M. Hruby et al., "The Evolution of Federally Funded Research & Development Centers," *Public Interest Report,* Federation of American Scientists (Spring 2011), https://fas.org/pubs /pir/2011spring/FFRDCs.pdf.

71. Kevin R. Kosar, "The Quasi Government: Hybrid Organizations with Both Government and Private Sector Legal Characteristics," Congressional Research Service Report, RL 30533, June 22, 2011.

72. See Jack Newsham, "Boston Gets $1.35 Million Grant for 'Housing Innovation Lab,'" *Boston Globe,* December 15, 2014, http://www .bostonglobe.com/business/2014/12/15/boston-gets-bloomberg-grant -research-housing-innovation/gL5dUGMGuiFr5G3TvIMWXL/story .html.

7. Bringing Smarter Governance to Life

1. Elliot A. Rosen, *Hoover, Roosevelt, and the Brains Trust: From Depression to New Deal* (New York: Columbia University Press, 1977), 115–211.

2. Arthur M. Schlesinger, *The Age of Roosevelt: The Crisis of the Old Order, 1919–1933* (New York: Mariner Press, 1957), 398. See also

Schlesinger, *The Age of Roosevelt: The Coming of the New Deal, 1933–1935* (New York: Mariner Press, 1958).

3. "2014 Education Policy Rankings," *U.S. News and World Report,* http://grad-schools.usnews.rankingsandreviews.com/best-graduate -schools/top-education-schools/education-policy-rankings?int =9a2b08.

4. David J. Arkush, "Direct Republicanism in the Administrative Process," *George Washington Law Review* 81 (2013): 1458–1528. See also "Citizen Juries: A Radical Alternative for Social Research," *Social Research Update* 37, University of Surrey (Summer 2002), http://sru .soc.surrey.ac.uk/SRU37.html.

5. Reeve T. Bull, "Making the Administrative State 'Safe for Democracy': A Theoretical and Practical Analysis of Citizen Participation in Agency Decisionmaking," *Administrative Law Review* 65 (2013): 640–47.

6. Beth Simone Noveck, *Wiki Government: How Technology Can Make Government Better, Democracy Stronger, and Citizens More Powerful* (Washington, DC: Brookings Institution Press, 2009), 180.

7. Julia Metz, "EU Commission Expert Groups: Between Inclusive and Effective Policy-making," in *Expertise and Democracy,* ed. Cathrine Holst, ARENA Report 1/14 (Arena Centre for European Studies, University of Oslo, Norway, February 2014), 276, http://www.sv.uio.no /arena/english/research/publications/arena-publications/reports/2014 /report-01–14.pdf.

8. Kenneth Arrow et al., "The Promise of Prediction Markets," *Science* 16 (May 2008): 877–78.

9. *The Federal Advisory Committee Act of 1972, U.S. Code* 5 (1972), §§ 2, 5, 9–11.

10. Ibid., § 3; *Federal Advisory Committee Management Final Rule,* General Services Administration, C.F.R. 66 (2001), pt. 102–3.25.

11. *Federal Advisory Committee Act of 1972,* § 9.

12. *Federal Advisory Committee Management Final Rule,* pt. 102–3.30.

13. *Public Citizen v. Department of Justice,* 491 U.S. 440, 454 (1989).

14. U.S. General Services Administration, "When Is the Federal Advisory Committee Act (FACA) Applicable?" http://www.gsa.gov/portal /content/100794; *Federal Advisory Committee Management Final Rule,* pt. 102–3.40(e).

15. Rebecca J. Long and Thomas C. Beierle, "The Federal Advisory Committee Act and Public Participation in Environmental Policy," *Resources for the Future* (January 1999), http://www.rff.org /Documents/RFF-DP-99-17.pdf.

16. Laura Denardis and Mark Raymond, "Thinking Clearly about Multistakeholder Internet Governance" (paper presented at the Eighth Annual GigaNet Symposium, Bali, Indonesia, October 21, 2013).

17. Casey Coleman, "GSA Social Media Handbook" (U.S. General Services Administration, July 2009), http://www.gsa.gov/graphics /staffoffices/socialmediahandbook.pdf.

18. *Alabama-Tombigbee Rivers Coalition v. Department of Interior,* 26 F.3d 1103 (11th Cir. 1994).

19. *Nader v. Baroody,* 396 F. Supp. 1231 (D.C. 1975).

20. Croley, "Practical Guidance," 484 (citing *Food Chemicals, Inc. v. Davis* (D.C. 1974)).

21. *Food Chemical News, Inc. v. Davis,* 378 F. Supp. 1048 (Dist. Court, Dist. of Columbia 1974).

22. Administrative Conference of the United States, "The Federal Advisory Committee Act—Issues and Proposed Reforms: Committee on Collaborative Governance Proposed Recommendation," December 8–9, 2011, 2, http://www.acus.gov/sites/default/files /documents/Proposed-FACA-Recommendation-with-Amendments -12-6-20111.pdf.

23. General Services Administration, "When Is FACA Applicable?" pt. 102–3.35.

24. Ibid, pt. 102–3.150.

25. *The Federal Register Act of 1935, U.S. Code* 44 (1935), §§ 1501–1511.

26. Amy Bunk, "Federal Register 101," *Integrity Action* (Spring 2010), https://www.federalregister.gov/uploads/2011/01/fr_101.pdf.

27. *Federal Advisory Committee Act,* House Report No. 92–1017, Cong. Record vol. 118 (1972), 3491ff, http://www.gsa.gov/graphics/ogp /FACALegislationHistory1972.pdf.

28. General Services Administration, "When Is FACA Applicable?" pt. 102–3.150.

29. *The Federal Advisory Committee Act of 1972,* § 10.

30. Steven P. Croley and William T. Funk, "The Federal Advisory Committee Act and Good Government," *Yale Journal on Regulation* 14 (1997): 464.

31. *Center for Auto Safety v. Cox*, 580 F.2d 689, 694 (D.C. Cir. 1978).

32. U.S. General Services Administration, "FACA 101," http://www.gsa .gov/portal/content/244333.

33. *Federal Advisory Committee Act of 1972,* § 13.

34. Philip Pettit, *Republicanism: A Theory of Freedom and Government* (Oxford, UK: Oxford University Press, 1999), 140.

35. John Ferejohn, "Conclusion: The Citizens' Assembly Model," in *Designing Deliberative Democracy: The British Columbia Citizens' Assembly,* ed. Hilary Pearse and Mark E. Warren (Cambridge, UK: Cambridge University Press, 2008), 209.

36. Administrative Conference of the United States, "The Federal Advisory Committee Act—Issues and Proposed Reforms: Committee on Collaborative Governance Proposed Recommendation," December 8–9, 2011, http://www.acus.gov/sites/default/files/documents /Proposed-FACA-Recommendation-with-Amendments-12–6– 20111.pdf.

37. Jill M. Hruby, Dawn K. Manley, Ronald E. Stoltz, Erik K. Webb, and Joan B. Woodard, "The Evolution of Federally Funded Research & Development Centers," *Public Interest Report* (Spring 2011): 24–30.

38. Science and Technology Policy Institute website, https://www.ida.org /stpi.php.

39. Schlesinger, *The Age of Roosevelt,* 420; and Rosen, *Hoover, Roosevelt, and the Brains Trust,* 160–61.

8. Smarter Citizenship

1. Frank Fischer, *Citizens, Experts and the Environment: The Politics of Local Knowledge* (Durham, NC: Duke University Press, 2000), 16.

2. Sheila Jasanoff, *The Fifth Branch: Science Advisers as Policymakers* (Cambridge, MA: Harvard University Press, 1990).

3. Harry Collins, *Are We All Scientific Experts Now?* (New York: Polity Press, 2014).

4. Harry Collins, "The Third Wave of Science Studies: Developments and Politics," *Japan Journal for Science, Technology & Society* 20

(2011), http://www.cardiff.ac.uk/socsi/contactsandpeople/harrycollins
/expertise-project/preprints/.

5. Collins, *Are We All Scientific Experts Now?* Kindle ed., location 216.

6. Sheila Jasanoff, "Quality Control and Peer Review in Advisory
Science," in *The Politics of Scientific Advice: Institutional Design for
Quality Assurance,* ed. Justus Lentsch and Peter Weingart (Cam-
bridge, UK: Cambridge University Press), 19.

7. William J. Sutherland et al., "A Collaboratively-Derived Science-
Policy Research Agenda," *PLoSOne* (March 9, 2012), http://www
.plosone.org/article/info%3Adoi%2F10.1371%2Fjournal.pone.0031824.

8. K. Sabeel Rahman, "Governing the Economy: Markets, Experts, and
Citizens," (PhD diss., Department of Government, Harvard University, 2013), 9, 11.

9. Center for Responsive Politics, OpenSecrets.org, "Influence and
Lobbying: Top Spenders," https://www.opensecrets.org/lobby/top.php
?showYear=2014&indexType=i.

10. Stephen P. Turner, *Liberal Democracy 3.0: Civil Society in an Age of
Experts* (London: Sage, 2003), 337.

11. Cathrine Holst and Anders Molander, "Epistemic Democracy and the
Accountability of Experts," in *Expertise and Democracy,* ed. Cathrine
Holst, ARENA Report 1/14 (Arena Centre for European Studies,
University of Oslo, Norway: February 2014), http://www.sv.uio.no
/arena/english/research/projects/episto/index.html.

12. Harry Collins and Charles Evans, *Rethinking Expertise* (Chicago:
University of Chicago Press, 2007), 8.

13. Paul Erickson et al., *How Reason Almost Lost Its Mind: The Strange
Career of Cold War Rationality* (Chicago: University of Chicago Press,
2013), 5.

14. Robert Geyer and Samir Rihani, *Complexity and Public Policy: A New
Approach to 21st Century Politics, Policy and Society* (Oxford, UK:
Routledge, 2010), 15.

15. Benjamin Barber, *Strong Democracy: Participatory Politics for a New
Age,* (Berkeley, CA: University of California Press, 1984), 47.

16. Erickson et al., *How Reason Almost Lost Its Mind,* 4.

17. Geyer and Rihani, *Complexity and Public Policy,* 61.

18. Philip E. Tetlock, *Expert Political Judgment: How Good Is It? How Can
We Know?* (Princeton, NJ: Princeton University Press, 2005), 20.

19. Philip E. Tetlock, "How to Win at Forecasting," in *Thinking: The New Science of Decision-Making, Problem-Solving and Prediction,* ed. John Brockman (New York: Harper, 2013), 18–39, https://edge.org /conversation/how-to-win-at-forecasting.

20. Braden R. Allenby and Daniel Sarewitz, *The Techno-Human Condition* (Cambridge, MA: MIT Press, 2011), 90.

21. Daniel Sarewitz, "Looking for Quality in All the Wrong Places," in *The Politics of Scientific Advice: Institutional Design for Quality Assurance,* ed. Justus Lentsch and Peter Weingart (Cambridge, UK: Cambridge University Press), 69.

22. Anthony Giddens, *Beyond Left and Right: The Future of Radical Politics* (Palo Alto, CA: Stanford University Press, 1994), 4.

23. Dr. Mark Wilson, email to the author, December 16, 2014.

24. National Lab Day, http://www.whitehouse.gov/open/innovations /national-lab-day.

25. Torfaen Local Services Board, *Wisdom, Wealth and Well-Being: New Opportunities for a Connected Wales,* Report, October 2011, http://www.torfaen.gov.uk/en/Related-Documents/Wisdom-Wealth -and-Well-being/Wisdom-Wealth-and-Wellbeing-Phase-1-Findings .pdf.

26. Brook Manville and Josiah Ober, *A Company of Citizens: What the World's First Democracy Teaches Leaders about Creating Great Organizations* (Boston, MA: Harvard Business Review Press, 2003).

27. Josiah Ober, *Democracy and Knowledge: Innovation and Learning in Classical Athens* (Princeton, NJ: Princeton University Press, 2009), 133.

28. Ibid., 95.

29. Ibid., 208.

30. Ibid., 38.

31. Daron Acemoglu and James Robinson, *Why Nations Fail: The Origins of Power, Prosperity and Poverty* (New York: Crown, 2012).

32. Ober, *Democracy and Knowledge,* 1.

33. Peter D. Blair, "Scientific Advice for Policy in the United States: Lessons from the National Academies and the Former Congressional Office of Technology Assessment," in Lentsch and Weingart, *Politics of Scientific Advice,* 310.

34. Paul Light, *Government by Investigation: Congress, Presidents, and the Search for Answers, 1945–2012* (Washington, DC: Brookings Institution Press, 2014), 65–66.

35. Blum Center, POV Video, *Can Experts Solve Poverty?* with Khalid Kadir, http://blumcenter.berkeley.edu/globalpov/.

36. Timothy Mitchell, *Rule of Experts: Egypt, Techno-Politics, Modernity* (Berkeley: University of California Press, 2002).

37. Norman Ornstein and Michael Mann, *It's Even Worse Than It Looks: How the American Constitutional System Collided with the New Politics of Extremism* (New York: Basic Books, 2012).

38. Lentsch and Weingart, *Politics of Scientific Advice,* 8.

Conclusion

Epigraph: Benjamin Barber, *Strong Democracy: Participatory Politics for a New Age,* (Berkeley: University of California Press, 1984), 311.

1. Kath Xu, "Admissions Says Yes to 9 Percent of Early Applicants," *The Tech,* January 8, 2014, http://tech.mit.edu/V133/N61/earlyaction.html.

2. Douglas Thomas and John Seely Brown, *A New Culture of Learning: Cultivating the Imagination for a World of Constant Change,* 1st ed. (CreateSpace Independent Publishing Platform, January 4, 2011).

3. Thomas L. Friedman, "How to Get a Job at Google," *New York Times,* February 22, 2014, http://www.nytimes.com/2014/02/23/opinion /sunday/friedman-how-to-get-a-job-at-google.html?-r=0.

4. Claudia Goldin and Lawrence F. Katz, *The Race between Education and Technology* (Cambridge, MA: Belknap Press, 2008), 5.

5. Suzanne Mettler, "College, the Great Unleveler," *New York Times,* http://opinionator.blogs.nytimes.com/2014/03/01/college-the-great -unleveler/?_php=true&_type=blogs&_r=0.

6. Clay Shirky, "Your Massively Open Offline College Is Broken," *The Awl,* February 7, 2013, http://www.theawl.com/2013/02/how-to-save -college.

7. Matthew Taylor, "The Power to Create," a speech delivered July 8, 2014, http://www.thersa.org/events/audio-and-past-events/2014/the -power-to-create.

8. Danielle Allen, *Our Declaration: A Reading of the Declaration of Independence in Defense of Equality* (New York: Liveright, 2014), 258.

9. Gov. Ted Strickland, "I Tried to Live on Minimum Wage for a Week," *Politico,* July 27, 2014, http://www.politico.com/magazine/story/2014/07/a-mile-in-shoes-of-the-minimum-wage-worker-109418.html.

10. Justus Lentsch and Peter Weingart, "Introduction: The Quest for Quality as a Challenge to Scientific Policy Advice: An Overdue Debate," in *The Politics of Scientific Advice: Institutional Design for Quality Assurance,* ed. Justus Lentsch and Peter Weingart (Cambridge, UK: Cambridge University Press), 4.

Acknowledgments

My son, Amedeo Max, was born the same year as his "older brother," the White House Open Government Initiative. Since that time, he has grown into a marvel and a wonder of a human being. He has been my wise, patient, and funny companion—generously dispensing "battery charges"—during the far-flung travel, intensive writing sprints, and slow editing slog that went into making this book. He is already full of the innate problem-solving ability and the willingness to give that make him tomorrow's paradigmatic smarter citizen. It is to him that this book and the future it optimistically portends are dedicated.

Good editors are hard to come by, and I am indebted to two at Harvard University Press. First, Elizabeth Knoll, who accepted this book based on a long email and a longer lunch. When Elizabeth moved from the Press to the University, Brian Distelberg inherited the mantle of supporter and cheerleader. He did the heavy lifting that made a good draft into a much better book. I am deeply grateful to Monroe Price and Jack Balkin, who—initially as anonymous commenters and later as much-welcomed mentors—did precisely what an author would wish for, namely recommend publication enthusiastically while providing detailed, constructive, and insightful suggestions. Copy editor Janet Mowery and Brian Ostrander of Westchester Publishing Services lavished attentive care on the manuscript. David Dembo was a critical and careful reader and a cherished friend throughout, who supported this project with excellent proofreading and even better cooking and conversation.

Alan Kantrow—friend and trusted advisor for twenty years—perused every chapter, sometimes more than once. To the extent the book reads well, he deserves the credit for pushing me to extract a compelling narrative out of once academic prose.

Alan is one of my colleagues at the Governance Lab, a think-and-do-tank, which studies how to use technology to improve how we govern. The themes of this book are closely intertwined with the work we do with Gov-Lab's collaborating partners, governments, nonprofits, and other organizations that are willing to experiment with doing things differently and measuring what works. At GovLab, I have the good fortune to work with people whom I love but also admire for their belief that we can govern in ways that are more legitimate and more effective and for their passionate commitment to doing good in the world. Thank you to all the GovLabbers: Audrey Ariss, Sean Brooks, Dinorah Cantu-Pedraza, Robyn Caplan, Emma Clippinger, Luis Daniel, Antony DeClerq, Fred DeJohn, Cosmo Fujiyama, Catherine Garcia, Seena Ghaznavi, Jessica Goldfin, Samantha Grassle, Joel Gurin, Celia Hannon, Kevin Hansen, Erika Harris, Maria Hermosilla, John Krauss, Laura Manley, Claudio Mendonca, Andrew Miller, Celia Parraud-Apparu, Claudia Paz, Shankar Prasad, Christina Rogawski, Julia Root, David Sangokoya, Batu Sayici, Shruti Shannon, Anupama Sharma, Manik Suri, Irene Tello, Basilio Valdehuesa, Lauren Yu, and Nikki Zeichner. All of them deeply enriched the writing process with their intellectual energy and enthusiasm. Arnaud Sahuguet, Andrew Young, and Jill Raines were especially active collaborators. My GovLab co-founder Stefaan Verhulst curates the weekly *GovLab Digest,* the single best source of news about innovations in governance. Many stories in this book are from sources he generously shared.

Smart Citizen, Smarter State is an exploration across disciplines of the advances in technology and the social sciences that are enabling more open, expert, and participatory ways of doing and deciding. It brings together insights from several literatures relevant to governance, including administrative and information law; the socio-cultural attributes of decisionmaking studied in policy sciences; developments in computer, information, and web sciences as they relate to designing improved processes for information exchange between citizens and institutions; and the philosophy and sociology of expertise. I could not have begun to master any, let alone all, of these fields without close collaboration from friends and colleagues whose wisdom is sprinkled throughout: Frank Baitman, Barak Berkowitz, Sir Tim Berners Lee, Mike Bracken, Herbert Burkert, Fadi Chehade, Aneesh Chopra, Laurence Claus, Norm Eisen, Francois Grey, David Ferriero, Jim Fishkin, Anindya Ghose, Danny Goroff, Timo Hannay, Jim

Hendler, Cesar Hidalgo, Panos Ipeirotis, Joi Ito, David Johnson, Steven Berlin Johnson, Tom Kalil, Akash Kapur, Tim Kelsey, Stephen Kosslyn, Vivek Kundra, Karim Lakhani, Julia Lane, Peter Levin, Natalia Levina, Raphael Majma, Carl Malamud (thanks for the Leonard White), Paul Maltby, Richard Matasar, Andrew McLaughlin, Steve Midgley, Geoff Mulgan, Tiago Peixoto, Sandy Pentland, Frank Petito, Ken Prewitt, Richard Price, Sabeel Rahman, Rakesh Rajani, Hollie Rousson-Gilman, Deb Roy, David Schoenbrod, Sir Nigel Shadbolt, Sonal Shah, Richard Sherwin, Rohan Silva, Katepalli Sreenivasan, Fred Turner, Eileen Twiggs, Mark Wilson, Chris Wong, and Hooman Yaghoobzadeh. Special thanks to Robyn Sturm-Steffen, the other half of the original White House Open Government Initiative. I am indebted to them as much for their friendship and good cheer as for their sound advice. Sadly, Richard Hackman passed away prematurely. He was a great mentor and a gentle soul. I miss him.

My late father, Simon Noveck, passed along his lifelong passion for democratic theory and practice to me at an early age. My mother, Doris Noveck, continues to pass along the most relevant and useful news items and insights. Her intellectual curiosity is contagious. Without her, I would know far less about the world. She is the ur-citizen-expert. Thank you, too, to my family—Novecks, Dembos, and Viezels—for their encouragement and patience.

This book benefited tremendously from the support of deans and colleagues at New York Law School, New York University, and MIT. Within the scholarly community, I am especially grateful to the members of the MacArthur Network on Opening Governance, who come together three times each year to workshop papers and projects like this book, and to my extended academic family at the Yale Law School Information Society Project. Several organizations provided the backing to enable the research and projects described in this book and to free up the time to write it, including the John D. and Catherine T. MacArthur Foundation, the John S. and James L. Knight Foundation, and Google.org.

Even the most solitary work is a highly social enterprise. The growing ranks of the open government community, whose public-minded and entrepreneurial members number in the tens of thousands, are the real inspiration for the belief that people are capable and, if we only knew how to ask them, they would willingly serve.

Index

Big data, 15, 100, 109
Bing, 139
Biosensors International, 151
Blake, William, 27
Blogging, xiv
Bock, Laszlo, 269
Boko Haram, 18
Bolden, Charles, 251
Boorstin, Daniel, 58
Boston University, 8
Bourdieu, Pierre, 49
Boyte, Harry, 78
Brahmachari, Samir, 5
Brain Trusts: FDR's, 209–210; modern,
    210–213; expanded to private universi-
    ties, 213–214; as directory of directo-
    ries, 215–218; matching supply with
    demand in, 218–222; experiments with,
    222–226; Federal Advisory Com-
    mittee Act and, 226–233; new legal
    framework for, 237–240; governance
    and, 254–255. *See also* Expertise
Brandeis, Louis, 144
Brazil, 168, 281–282n41
Breen, T. H., 44
Brown, Mark, 200
Building inspections, 35
Bull, Reeve, 220–221
Bureaucracies: participatory, 19–21;
    professionalization of, 50, 60, 72–74;
    development of, 53–54; professions
    distinguished from, 87; alternatives to,
    92; representative democracy and,
    300n54
Bureau of the Budget (U.S.), 187, 189
Bush, George W., 244
Bush administration (G.W. Bush):
    federal websites under, xiii; environ-
    mental advisory committees under,
    205–206

Camouflage military uniforms, 33–34
Canadian Open Government Directive, 14
Canetti, Elias, 84
Caplan, Bryan, 81
Cardiopulmonary resuscitation (CPR), 39
Carnegie, Andrew, 63
Castronova, Ted, 142–143
Catholic Church, 7
Cell phones, 75–76, 181–183
Center for Devices and Radiological
    Health (CDRH), 150, 153, 154, 178
Center for Global Development, 21
Central Falls (Rhode Island), 13
Certifications, 120–124. *See also*
    Credentials
Challenge.gov, 9, 24, 196
Chan Lai Fung, 213
Charismatic leadership, 89–90
Charter on Open Data, 264
Chesbrough, Henry W., 2, 91
China, 167; mobile advertising in, 182
Churchill, Winston, 241
Citations, 124, 126
CiteSeer, 124
Citizen's Briefing Book, ix–x, 285n74
Citizen science, 4–6
Citizenship: expertise versus, 241;
    transformed by smarter governance,
    250–259
Civic Science, 6
Civil service: professionalization and,
    28–29, 47–50, 53, 70–71; better use of,
    40; current number of employees in,
    46; development of bureaucracy in,
    53–54; fixed salaries for, 60; nine-
    teenth century growth of, 67–70
Clark, William, 44
Clay, Henry, 45
Cleisthenes, 255
Clinton, Bill, 198

International City/County Management Association, 252
Internet: used by Obama campaign, ix; protests launched on, 18; unreliability of information flow on, 31; shifts in political opinion on, 86; nonhierarchical flow of information on, 91; of Things, 100; tracking browsing on, 110–111, 304–305n36; expertise and, 111–113; as permanent record, 128; online meetings on, 236; for identifying expertise, 268
Invisible Children (organization), 18
Ipeirotis, Panos G., 115
ITalki (expertise language platform), 105

Janis, Irving, 85
Jasanoff, Sheila, 199–200, 241
Jefferson, Thomas, 44, 45, 80, 184
Juries, 220–221, 264
Justice, Department of (U.S.), 233–234

Kaggle (firm), 10, 123
Kakfa, Franz, 56
Kalil, Thomas A., 146
Kamen, Dean, 130
Kennedy, John F., 189
Kenya, 23
Kere, Diebedo Francis, xv
Khan Academy (learning platform), 113
Kickstarter, 170
Klout, 118–119
Knowledge Network, 252
Knowledge networks, 101. *See also* Expert networking
Kony, Joseph, 18

Labor, Department of (U.S.), 204
Lakhani, Karim, 163, 180

Landemore, Helene, 96, 98
Land grant universities, 64
Langdell, Christopher Columbus, 64
Laski, Harold, 27, 72, 77
Law, study of, 64, 71
Lawyers, 61
Leadbeater, Charles, 179
Le Bon, Gustave, 66
Legislation: Paperwork Reduction Act, xii–xiii, 185, 186, 192–200; Administrative Procedure Act, 19, 185, 186; on health care reform, 34; Food and Drug Administration Safety Innovation Act, 156; on federal agencies and advisory committees, 185–186, 189–192; Freedom of Information Act, 190, 235; needed reforms in, 237–240. *See also* Federal Advisory Committee Act
Leona Helmsley Trust, 163
Levina, Natalia, 132
Lewin, Kurt, 84
Lewis, Meriwether, 44
Light, Paul, 260
Lincoln, Abraham, 45
LinkedIn: matching volunteers with jobs, 110; recommendations on, 113, 127; forms on, 116, 120; to identify Brain Trust members, 216
Linux (operating system), 270
Lippmann, Walter, 75
Lipset, Seymour Martin, 83–84
Liquid Feedback platform, 13, 168
Literature, perceptions of time in, 56
LiveNinja, 134
Local Motors (firm), 3
Lowi, Theodore, 70
Lumosity, 124
Lynda 112, 113
Lysenko, Trofim D., 243